NCRP REPORT No. 96

COMPARATIVE CARCINOGENICITY OF IONIZING RADIATION AND CHEMICALS

Recommendations of the
NATIONAL COUNCIL ON RADIATION
PROTECTION AND MEASUREMENTS

Issued March 1, 1989

RC 268.65
N 35
1989

National Council on Radiation Protection and Measurements
7910 WOODMONT AVENUE / Bethesda, MD 20814

LEGAL NOTICE

This report was prepared by the National Council on Radiation Protection and Measurements (NCRP). The Council strives to provide accurate, complete and useful information in its reports. However, neither the NCRP, the members of NCRP, other persons contributing to or assisting in the preparation of this report, nor any person acting on the behalf of any of these parties (a) makes any warranty or representation, express or implied, with respect to the accuracy, completeness or usefulness of the information contained in this report, or that the use of any information, method or process disclosed in this report may not infringe on privately owned rights; or (b) assumes any liability with respect to the use of, or for damages resulting from the use of any information, method or process disclosed in this report, *under the Civil Rights Act of 1964, Section 701 et seq. as amended 42 U.S.C. Section 2000e et seq. (Title VII) or any other statutory or common law theory governing liability.*

Library of Congress Cataloging-in-Publication Data

National Council on Radiation Protection and Measurements.
   Comparative Carcinogenicity of Ionizing Radiation and Chemicals.

   (NCRP Report ; No. 96)
   Bibliography: p.
   Includes index.
   1. Carcinogenicity testing. 2. Tumors, radiation induced.
   3. Ionizing radiation — Health aspects. 4. Health risk assessment.
   I. Title. II. Series.
   RC268.65.N35   1989   616.99'40741   88-15200
   ISBN 0-913392-96-0

Copyright © National Council on Radiation
Protection and Measurements 1989
All rights reserved. This publication is protected by copyright. No part of this publication may be reproduced in any form or by any means, including photocopying or utilized by an information storage and retrieval system without written permission from the copyright owner, except for brief quotation in critical articles or reviews.

Preface

This Report reviews the methodologies used to evaluate the carcinogenic risks of radiation, with reference to their applicability for evaluating the carcinogenic risks of chemicals. Where possible, the mechanisms of action and dose-response relationships of ionizing radiation are compared with those of chemicals. In a few instances it has been possible to compare mathematical models used to predict the carcinogenic risks of radiation with those used to compare the carcinogenic risk of chemicals, and to provide a gross evaluation of the usefulness of applying some of the methods currently used in expressing the risk of radiation exposure to the expression of risks from chemical exposure.

This Report was prepared by NCRP's Task Group on the Comparative Carcinogenicity of Pollutant Chemicals, under the Study Group on Comparative Risk. Serving on the Task Group were:

A.C. Upton, Chairman
New York University Medical Center
Institute of Environmental Medicine
Tuxedo, New York

R.E. Albert
Department of Environmental Health
University of Cincinnati
Cincinnati, Ohio

G.W. Beebe
Clinical Epidemiology Branch
National Cancer Institute
Bethesda, Maryland

A.V. Ray
Medical Research Laboratory
Pfizer, Inc.
Groton, Connecticut

R. Wilson
Lyman Laboratory
Harvard University
Cambridge, Massachusetts

J.C. Barrett
Environmental Carcinogenesis Group
National Institutes of Health
Research Triangle Park,
North Carolina

D.W. Nebert
Developmental Pharmacology Branch
National Institute of Child Health and
 Human Development
Bethesda, Maryland

R.R. Tice
Medical Department
Brookhaven National Laboratory
Upton, New York

S.H. Yuspa
National Cancer Institute
Bethesda, Maryland

Serving on the Study Group on Comparative Risk were:

W.K. Sinclair, Chairman
National Council on Radiation Protection and Measurements
Bethesda, Maryland

V.P. Bond
Medical Department
Brookhaven National Laboratory
Upton, New York

T.S. Ely
Pittsford, New York

D.W. Moeller
Harvard School of Public Health
Boston, Massachusetts

NCRP Secretariat — **W.M. Beckner**

The Council wishes to express its gratitude to the members of the Task Group and the Study Group for the time and effort devoted to the preparation of this report.

Bethesda, Maryland
December 15, 1988

Warren K. Sinclair
President, NCRP

Contents

Preface	iii
1. Introduction	1
2. Historical Background	2
2.1 Evolution of Approaches for Assessing the Carcinogenic Risks of Ionizing Radiation	2
2.2 Evolution of Approaches for Assessing the Carcinogenic Risks of Chemicals	3
2.3 Role and Activities of NCRP in Risk Assessement	3
2.4 Objectives and Scope of this Report	4
3. Comparative Nature and Mechanisms of Carcinogenic Effects of Radiation and Chemicals	5
3.1 Unicellular, Monoclonal Origin of Cancer	5
3.2 Multi-causal, Multi-stage Nature of Carcinogenesis (Initiation, Co-carcinogenesis, Promotion, Progression)	6
3.3 Reparative and Homeostatic Processes (Anti-carcinogenesis)	11
3.3.1 Susceptibility in Relation to Age	11
3.3.2 Role of Immune System	12
3.3.3 Influence of Genetic Background	12
3.3.4 Anti-carcinogens	12
3.4 Dose-Response Relationships	13
4. Nature, Distribution, and Sources of Ionizing Radiation in the Environment	18
4.1 Physical Properties of Ionizing Radiation	18
4.2 Sources and Levels in the External and Internal Environment	18
4.3 Assessment of Exposure to Ionizing Radiation	19
4.3.1 Introduction	19
4.3.2 Route of Exposure and Distribution of Absorbed Dose	21
4.3.3 Metabolism: Influence on Distribution of Radionuclides in the Body	22
4.4 Summary	22
5. Nature, Distribution, and Sources of Carcinogenic Chemicals in the Environment	24
5.1 Chemical Properties	24

CONTENTS

- 5.2 Sources and Levels of Carcinogenic Chemicals in the External Environment and in Various Tissues of the Body 24
- 5.3 Assessment of Exposure 31
  - 5.3.1 Units of Measurement, Dosimetry 31
  - 5.3.2 Ambient Levels in the Environment 31
  - 5.3.3 Route of Exposure 33
  - 5.3.4 Concentrations in the Body 34
- 5.4 Pharmacokinetics 36
  - 5.4.1 Uptake 37
  - 5.4.2 Distribution 37
  - 5.4.3 Metabolism 37
  - 5.4.4 Excretion 43
  - 5.4.5 Route of Administration 43
- 5.5 Summary 46

6. Carcinogenic Effects of Radiation and Chemicals in Man ... 48

- 6.1 Radiation 48
  - 6.1.1 Characterization of Exposure 48
    - 6.1.1.1 Part of Body Exposed 48
    - 6.1.1.2 Dose to Organ or Tissue 49
    - 6.1.1.3 Dose Rate and Fractionation 50
    - 6.1.1.4 Quality of Radiation (Linear Energy Transfer) 51
  - 6.1.2 Dose-Response Relationships 52
  - 6.1.3 Host Factors 56
  - 6.1.4 Other Environmental Factors 59
  - 6.1.5 Organ Differences in Response 59
  - 6.1.6 Distribution of Radiogenic Tumors in Relation to Time After Exposure 60
  - 6.1.7 Relation of Radiogenic Cancer Incidence to Natural Incidence 61
  - 6.1.8 Formulation of Dose-Response Models of Radiation Carcinogenesis 63
- 6.2 Chemicals 68
  - 6.2.1 Route of Exposure and Site of Action 68
  - 6.2.2 Dose-Response Relationship 68
  - 6.2.3 Host Factors 72
  - 6.2.4 Distribution of Chemically Induced Cancers in Relation to Time after Exposure 73
  - 6.2.5 Relation to Other Environmental Factors and Natural Incidence 74
- 6.3 Summary 74

7. **Extrapolation From Laboratory Models to the Human** 76
   7.1 Effects on DNA, Genes, and Chromosomes 76
   7.2 Transforming Effects on Cells in Culture 88
       7.2.1 Introduction 88
       7.2.2 Systems to Study Neoplastic Transformation of Cells in Culture 89
       7.2.3 Detection of Carcinogens with Cell Transformation Assays 90
       7.2.4 Dose-Response Data for Transformation of Cells by Chemicals and Radiation 92
       7.2.5 Modifiers of Chemical- and Radiation-Induced Cell Transformation 94
       7.2.6 Mechanistic Studies of Cell Transformation 94
       7.2.7 Role of Oncogenes in Multi-stage Carcinogenesis .. 97
   7.3 Effects on Laboratory Animals 98
       7.3.1 Carcinogenesis in the Skin 98
       7.3.2 Carcinogenesis in the Liver 99
       7.3.3 Carcinogenesis in the Respiratory Tract 100
       7.3.4 Carcinogenesis in the Mammary Gland 101
       7.3.5 Carcinogenesis in the GastroIntestinal Tract 101
       7.3.6 Carcinogenesis in the Hematologic Organs 102
       7.3.7 Carcinogenesis in Other Organs 103
       7.3.8 Transplacental Carcinogenesis 103
   7.4 Summary 104
8. **Risk Assessment** 106
   8.1 Elements Involved in Risk Assessment 106
   8.2 Dose-Incidence Relations 108
       8.2.1 Ionizing Radiation 108
       8.2.2 Chemicals 111
   8.3 Age- and Time-Distribution of Neoplasms 119
   8.4 Application to Specific Agents or Exposure Situations .. 120
   8.5 Expression of Risk 120
   8.6 Uses of Risk Analyses 121
       8.6.1 Uses for Estimates of Risks of Radiation 121
       8.6.2 Uses of Estimates of Risks of Chemicals 122
       8.6.3 Risk Assessment for Attribution of Risk 126
   8.7 Summary 127
9. **Conclusions** 129
References 131
The NCRP 162
NCRP Publications 169
Index 178

1. Introduction

This Report was prepared for the purpose of evaluating the extent to which the principles and methods that have been developed for use in assessing the carcinogenic risks of ionizing radiation are applicable in assessing the carcinogenic risks of chemicals. In addressing this question, a review of the status of knowledge concerning the carcinogenic effects of radiation and chemicals, with particular reference to their comparative mechanisms of action and dose-response relationships, was performed. Also reviewed were mathematical models for predicting the carcinogenic risks of radiation and chemicals under exposure conditions where human data are fragmentary or lacking.

No attempt is made herein at an exhaustive presentation of the vast amount of pertinent information that is available. Instead, this Report seeks to summarize the cogent evidence and to provide references to authoritative sources for those readers desiring further details.

2. Historical Background

2.1 Evolution of Approaches for Assessing the Carcinogenic Risks of Ionizing Radiation

Injuries from large doses of ionizing radiation were recognized almost immediately after the discovery of x rays by Roentgen in 1895, but the possibility that there might be no threshold dose for some types of harmful effects was not considered seriously for almost half a century. By the 1950's, however, the observed linearity of the dose-response curve for mutagenesis in *drosophila* and certain microbial organisms led to abandonment of the view that a sharp threshold for such effects can be assumed to exist (NCRP, 1954; NAS/NRC, 1956; UNSCEAR, 1958). Later, the increased incidence of leukemia in patients treated by irradiation of the spine for ankylosing spondylitis (Court-Brown and Doll, 1957), in atomic-bomb survivors (Moloney and Kastenbaum, 1955), and in radiologists (March, 1944; Ulrich, 1946) led Lewis (1957) to postulate a linear nonthreshold dose-incidence relationship for this disease, and to suggest that 10-20 percent of leukemia in the general population might be attributable to natural background radiation. In the interim, although experimental and epidemiological evidence against the threshold hypothesis has become stronger, our understanding of low-dose-rate effects is not complete and thus, the magnitude of the risk for leukemia and other cancers at low dose must be inferred from data at higher doses and, therefore, remains a matter of uncertainty.

Recent efforts to assess the carcinogenic risks of low-level irradiation have utilized hypothetical dose-response models as a basis for extrapolation from epidemiological observations at higher doses and dose rates. Selection among possible alternative dose-response models has been guided by consideration of relevant radiobiological mechanisms of the effects in question. In spite of the fact that knowledge of the mechanisms is uncertain, the data have enabled the formulation of numerical risk estimates which have been helpful in radiological protection, in exploring the factors that influence the risk of radiogenic cancer, and in handling claims for compensation.

For better or worse, the approaches to risk assessment that have been developed for purposes of radiological protection have set a pattern which has influenced approaches for assessing the risks of chemical mutagens and carcinogens.

2.2 Evolution of Approaches for Assessing the Carcinogenic Risks of Chemicals

Certain chemicals, like radiation, have long been known to be capable of causing carcinogenic effects at high doses, but the possibility of their carcinogenicity at low doses has been considered only relatively recently. The hypothesis that the effects of carcinogenic chemicals may not show a sharp threshold has received support from the recognition that a high percentage of them possess mutagenic activity. Nevertheless, the diversity in chemical structure of known carcinogens and the evidence that carcinogenic effects may be mediated through different pathways—including initiating effects, promoting effects, cocarcinogenic effects and other mechanisms—have complicated the development of any unifying hypothesis with which to assess the carcinogenicity of all compounds. For most chemicals, moreover, pharmacokinetic variables complicate the relation between exposure and an effect on target cells.

In addition, while hundreds of chemicals have been found to cause carcinogenic effects in laboratory animals, definite evidence of human carcinogenicity exists for only a score or more (IARC, 1982). At present, therefore, estimates of the carcinogenicity of most chemicals for man must be based largely on tests in laboratory animals, supported by ancillary data from other types of bioassays and from knowledge of the relationship between molecular structure and biological activity.

2.3 Role and Activities of NCRP in Risk Assessment

The National Council on Radiation Protection and Measurements (NCRP) has been concerned from its inception with assessment of the risks of low-level irradiation. In recent years, this concern has prompted increasingly detailed efforts to quantify the influence of dose, dose rate, and linear energy transfer (LET) on the biomedical effects of radiation, with particular reference to the carcinogenic risks of low-level exposure (NCRP, 1980) and the work-in-progress of NCRP Scientific Committee 40 on LET and relative biological effectiveness.

In addition to ionizing radiation, other forms of radiation and other physical and chemical agents have also been evaluated by NCRP with respect to their potential health impacts, comparative risks, and protection criteria. Furthermore, because radiation in combination with other physical and chemical agents may exert additive, synergistic, or even antagonistic effects, NCRP has concerned itself increasingly with assessment of the effects of chemical agents and the extent to which they may modify the carcinogenic effects of ionizing radiation. This report is thus a logical outgrowth of these concerns.

2.4 Objectives and Scope of this Report

This Report considers the degree to which concepts and principles evolving out of experience in assessing the risks of ionizing radiation at low levels may be applicable to assessment of the carcinogenic risks of chemicals. In reviewing this subject, salient comparisons between radiation and chemicals are surveyed, with respect to such parameters as entry into the body, distribution in cells and tissues, biological effects, dosimetry, and dose-response relationships. An attempt is made, insofar as possible, to integrate knowledge derived from epidemiological studies of human populations with relevant experimental data.

3. Comparative Nature and Mechanisms of Carcinogenic Effects of Radiation and Chemicals

3.1 Unicellular, Monoclonal Origin of Cancer

The monoclonal nature of cancer has been suggested by studies of human as well as experimental tumors. Studies of human tumors have utilized the X-linked glucose-6-phosphate dehydrogenase (G-6-PD) marker. Two isozymes of G-6-PD exist, A and B, which are readily separable by electrophoresis. In female heterozygotes, individual cells are type A or B, due to inactivation of one X chromosome, while tissues are mosaics of the two cell types and thus contain both enzymes. The demonstration of a single enzyme type in cancer tissue from such mosaics is strong evidence in support of monoclonal growth. Of a large series of benign, premalignant, and malignant human tumor types studied in this way, all were of monoclonal origin (Fialkow, 1977). Only two consistent exceptions to these observations have been found. Hereditary neurofibromas in von-Recklinghausen disease and hereditary trichoepitheliomas have both enzyme types, suggesting different pathogenic mechanisms of these tumors.

Experimental confirmation of the monoclonal origin of chemically induced cancers have been obtained in chimeric mice which were mosaic for electrophoretic variants of glucose phosphate isomerase (Iannaccone et al., 1978). When benign or malignant tumors of epithelial or fibroblastic type were produced by subcutaneous injection or skin painting of carcinogens, virtually all of the tumors were of the single enzyme types. In interstrain chimeras, where one strain was more sensitive to skin carcinogenesis, the tumor isozymes again were monoclonal but were much more commonly from the sensitive strain, suggesting that susceptibility determinants reside within the target cell. Similar results have been obtained with other experimental tumor induction models using chemical agents (Ponder, 1980), but such studies have not been reported for radiation carcinogenesis.

Studies such as these, while pointing out similarities between human and experimental carcinogenesis, also suggest that the mechanisms involved in carcinogenesis are not of an "infectious" or spreading type and do not seem to involve recruitment of normal cells to become cancer cells within the tumor mass. The single-cell origin of tumors is most consistent with the somatic mutation hypothesis for cancer initiation. By virtue of their potent mutagenic activity, most carcinogenic chemicals and radiation would be likely causal factors in the production of somatic mutations.

3.2 Multi-causal, Multi-stage Nature of Carcinogenesis (Initiation, Co-carcinogenesis, Promotion, Progression)

Human and animal data suggest that cancer evolves through successive discrete stages. Metaplasia is a well recognized precursor lesion in cancer of the lung, cervix, and oral cavity; intestinalization often precedes gastric cancer; and ductal hyperplasia is considered a premalignant lesion in breast cancer. Likewise, benign tumors of the skin, colon, bladder, and thyroid are commonly precursors to malignant lesions, often with a long and stable latent period.

Experimentally, carcinogenesis in complex epithelia appears to occur through a multistage mechanism. In the skin, liver, esophagus, colon, bladder, mammary gland, respiratory tract, and stomach, the cancer induction process can be subdivided into three stages: initiation, promotion, and progression, e.g., Farber, (1984). Initiation can result from a single exposure to a genotoxic carcinogen and is irreversible. The permanently altered initiated cell and its progeny may never form a tumor or in any way be recognizable in the target tissue. Subsequent exposure to tumor promoters permits the neoplastic change to be expressed in initiated cells, with the result that tumors then develop. Progression of the tumors toward increasing malignancy may result from further stimulation. In contrast to initiation, promoters must be given repeatedly to be effective; the effects of a single exposure are reversible. In its simplest form, this relationship would give rise to an ultimate tumor yield that is dependent on the dose of both initiator and promoter. A number of agents have been found to possess both initiating and promoting properties if given in sufficient doses; these are termed "complete" carcinogens.

Multiple premalignant stages are apparent during chemical induction of skin and liver cancer (Farber, 1984), and multistage processes have been identified also in experimental lung (Yarita et al., 1978), breast (Medina, 1976), bladder (Hicks et al., 1975), and colon cancer (Thurnherr et al., 1973) as well. In chemically initiated mouse skin

(Yuspa and Morgan, 1981) and rat tracheal epithelium (Terzaghi and Nettesheim, 1979), preneoplastic cells have been recognized by altered growth properties *in vitro*. Similarly, preneoplastic cells have been reported to be identifiable in radiation-induced leukemia by transplanting cells from irradiated mice into appropriately conditioned recipients (Haran-Ghere, 1978; Boniver *et al.*, 1981; Hirashima *et al.*, 1982).

The immediate and irreversible nature of initiation, and the demonstration that most carcinogens are mutagens, have implicated a mutational mechanism for initiation. However, mammalian cells treated with chemical carcinogens have a higher frequency of neoplastic transformation than of gene mutation at any one locus (Huberman *et al.*, 1976; Parodi and Brambilla, 1977), suggesting multiple oncogenic sites, damage of the genome at sites less likely to be repaired, e.g., tandem repeats, or genetic damage other than point mutations. A mutational mechanism is supported by the findings that: (1) transformation frequency increases if exposure to a chemical carcinogen occurs just prior to, or during, the DNA synthetic phase of the cell cycle (Bertram and Heidelberger, 1974), (2) DNA to which a chemical carcinogen is bound can serve as a template for DNA replication (Bates *et al.*, 1970), and (3) DNA synthesis and consequent cell division after carcinogen exposure stabilizes the potential for neoplastic change (Kakunaga, 1978). Whatever the precise mechanism, DNA is considered by many to be the critical cellular target for chemical initiators, and damage to DNA by carcinogens and repair of this damage have been studied extensively. In general, in a given biological system, the carcinogenic potency of an initiating chemical correlates with the extent to which the chemical can bind covalently to DNA, and with the nature of the products formed by this interaction.

DNA damage produced by chemicals and by radiation include single-strand and double-strand breaks, cross-links, various alterations in sugar and base moieties, base substitutions, and recombinational changes. Chromosomal aberrations are also observed. In general, dose-response studies are consistent with the interpretation that the carcinogenic action increases with increasing DNA damage up to the point where cell death or other excessive injury becomes predominant (Upton, 1982).

The specific interaction of chemicals or radiation with DNA activate cellular DNA repair processes which appear to play an important role in carcinogenesis. Eukaryotic cells show enhanced repair capacity for repair of viral nucleic acid if the host cells are first damaged by chemicals or irradiation. The enhanced susceptibility to cancer of patients with defective repair systems, such as those with xeroderma pigmentosum, suggests that intact repair mechanisms are protective against cancer development. In some cases, however, DNA repair may be error prone. Prokaryotic cells contain an inducible error prone recombinational repair system (Lehman and Bridges, 1977). Radiation exposure at levels that do not produce demonstrable mutagenic effects has also been shown to sensitize cells to the mutagenic action of other DNA damaging agents (Frank and Williams, 1982).

While considerable evidence supports a mutational mechanism for initiation, the genes involved are only beginning to be defined. Recently, DNA isolated from a variety of human tumors and transfected into certain cultured cells has caused the recipient cells to become transformed. In some cases, the genes responsible for the transformation, i.e., the oncogenes (see Section 7.2.7), have been identified as mutated derivatives of cellular genes (proto-oncogenes) which are homologous to transforming genes of retroviruses. Members of the *ras* gene family have been frequently identified. While the causal relationship between the mutation of a *ras* gene and the initiation of carcinogenesis is not firmly established, altered *ras* genes have been isolated from several experimental tumor sites at a very early stage in transformation (Balmain et al., 1984; Sukumar et al., 1983). Furthermore, activated *ras* genes have been isolated from mouse lymphomas induced by radiation or chemicals (Guererro et al., 1984). In this case, chemically induced lymphomas yielded transforming N-*ras* homologues while irradiation yielded k-*ras* homologues, suggesting that activation of one of several genes can lead to the same tumor phenotype.

Conversion of proto-oncogenes into active oncogenes capable of inducing a tumorigenic state appears to involve one of five processes: (1) over-expression resulting from the acquisition of a novel transcriptional promoter, (2) over-expression due to amplification, (3) addition of enhancer sequences, (4) deregulation, and (5) point mutations in the DNA sequence of the proto-oncogene (Land et al., 1983a). The presence of specific, nonrandom karyotypic alterations in human and animal cancers, and the nature of the chromosomal rearrangements suggest that gene transposition can alter gene expression, leading to a transformed state (Radman et al., 1982; Rowley, 1982, 1984; Yunis, 1983; Sandberg, 1983; Mitelman, 1984). Various kinds of ionizing radiations and the vast majority of chemical carcinogens are capable of causing the kinds of DNA damage that can lead to chromosomal rearrangements and thus, to oncogene activation.

Tumor promotion has been demonstrated in a number of experimental models, but the mechanism of promotion has been studied most extensively in mouse skin, rat liver, and cell culture. In skin, the most potent promoters are fatty acid esters of phorbol, a plant-derived diterpene alcohol. There is a strict structure-function relationship in the tumor-promoting potency of these compounds, which is determined in part by the length of the fatty acid ester groups, an optimum being 14 carbon lengths. The most potent of the phorbol esters is 12-0-tetradecanoyl phorbol-13-acetate. This substance is active in minute concentrations and for brief exposures, in vivo and in vitro, and induces biochemical changes (Blumberg, 1981). These changes include stimulation of macromolecular synthesis leading to hyperplasia, stimulation of polyamine synthesis, prostaglandin synthesis, protease production, alterations of cell membrane enzymes and glycoproteins, altered differentiation, and altered responses to other growth-controlling factors. These effects have been demonstrated in cells from a number of species, including man. The critical responses for tumor promotion have not yet been determined, but the pleiotropic effects induced by these agents at nanomolar concentrations and their discrete structure-activity relationships (Blumberg, 1981) have suggested a hormone-like action. There are specific cellular receptors for phorbol ester tumor promoters, and recent studies have identified a major receptor as protein kinase C (Blumberg et al., 1984). This kinase has been implicated in the specific effector activity of a number of hormones (Takai et al., 1982) and could mediate promoter-like activity in a variety of target organs. Tumor promoters appear to be tissue-specific but have in common the property that they induce proliferation in the target tissue. A list of promoting agents and their target tissues is shown in Table 3.1.

Promoting agents appear capable of enhancing radiation carcinogenesis in a manner similar to that in which they enhance chemical carcinogenesis. Thus, the rate of appearance and cumulative incidence of breast tumors are greatly increased by mammographic (Furth, 1975) or estrogenic (Shellbarger, 1981) stimulation after irradiation. Endocrine stimulation may also promote the pathogenesis of other radiation-induced tumors (Furth, 1975; Clifton and Sridharan, 1975), including those of the murine Harderian gland. Other examples of tumor promotion include the enhanced induction of thymic lymphomas in mice by administration of urethane after irradiation (Berenblum and Trainin, 1963), the enhanced induction of hepatocellular tumors in mice by administration of carbon tetrachloride after irradiation (Cole and Nowell, 1964), and the enhanced induction of kidney tumor in rats by contralateral nephrectomy after irradiation (Rosen and Cole, 1962).

The above are but a few examples of carcinogenesis in which the initiating effects of radiation have been enhanced by subsequent exposure to appropriate promoting stimuli. The cytotoxic effects of irradiation themselves have also been postulated to promote the development of radiation induced neoplasia in other situations; e.g., in the pathogen-

TABLE 3.1—*Examples of experimental promoting agents and their target tissues*

Tissues	Promoting Agents
Mouse skin	Phorbol esters (croton oil) Anthralin Cigarette smoke condensate Dihydroteleocidin (fungal product) 7-Bromomethylbenz[a]anthracene UV light Wounding
Rat liver	Phenobarbital DDT Polychlorinated biphenyl
Rat urinary bladder	Saccharin Cyclamate Tryptophan
Rat colon	Bile acids Cholestyramine High fat diets Wounding
Hamster lung	Chronic irritation (saline lavage)
Hormone-dependent tissues	Hormones

esis of osteogenic sarcomas in animals injected with radium-226 (Marshall and Groer, 1977). A situation in which radiation carcinogenesis may be inferred to involve mechanisms more nearly akin to those responsible for tumor promotion than to those responsible for tumor initiation pertains to those neolasia that are induced through indirect, or *abscopal*, effects. Examples include radiation-induced thymic lymphomas (Kaplan, 1967), ovarian tumors (Upton, 1967), and thyrotropic pituitary tumors (Furth, 1975) in mice.

A number of agents are known to act as co-carcinogens in chemical carcinogenesis. These chemicals are neither carcinogens nor are they necessarily promoters, but when given together with carcinogens they enhance the carcinogenic process (Van Duuren, 1976). Such agents abound in the environment and are of varied chemical structure. A number of phenols and aliphatic as well as aromatic hydrocarbons fall into this category. These agents may act by altering the uptake, distribution or metabolism of carcinogens or the susceptibility of the target cells or host. Multiple exposures to several initiators may produce a variety of results. In hamsters, combined treatments with benzo[a]pyrene administered intratracheally and N-nitrosodiethylamine administered systemically, at doses at which the individual carcinogens induce few or no malignant respiratory tract tumors, produce a high percentage of malignant tumors of the trachea and bronchi (Montesano et al., 1973). Chronic simultaneous feeding of four liver carcinogens, each at doses that produce no observed liver changes, can lead to the development of many liver cancers (Schmaehl, 1970). However, synergism is not always found. Simultaneous intratracheal instillation of benzo[a]pyrene and polonium-210 in hamsters leads to a respiratory cancer incidence which is equal to the sum of the incidences attributable to each individual exposure (Little et al., 1978). Antagonistic interactions also occur, i.e., simultaneous initiation of mouse skin with the potent carcinogen 7, 12-dimethylbenz[a]anthracene or 3-methylcholanthrene and the weak carcinogen benzanthracene or dibenzo[a,c]anthracene effectively inhibits the strong initiator (Slaga and Boutwell, 1977). This inhibition apparently is mediated by the induction of metabolic pathways leading to noncarcinogenic metabolites of the stronger agent.

Modification of chemical carcinogenesis by factors that alter the state of the target cell has been observed. Cells that are proliferating at the time of carcinogen exposure are particularly sensitive to carcinogenesis. Thus, induction of proliferation by chemicals or surgical procedures can markedly enhance initiation in experimental skin, colon, and liver carcinogenesis (Hennings et al., 1973; Pozharisski, 1975; Wogan, 1976). This same phenomenon may be reflected in the sensitivity of younger animals to chemical carcinogens.

3.3 Reparative and Homeostatic Processes (Anti-Carcinogenesis)

3.3.1 *Susceptibility in Relation to Age*

In general, susceptibility to carcinogens appears to decrease with age after birth. Thus, neonates are susceptible to a single low dose of many chemicals that require chronic administration in adults. The yield of benign and malignant tumors decreases with increasing age at initiation and promotion in skin carcinogenesis (Van Duuren et al., 1975). Experiments utilizing skin grafted from young and old animals have suggested that the age of the target tissue rather than that of the host determines susceptibility (Ebbesen, 1977). Some carcinogenesis model systems have demonstrated a striking age specificity for tumor development. Female Sprague-Dawley rats are highly susceptible to mammary cancer when exposed to 7, 12-dimethylbenz[a]anthracene at 50 days of age but are less sensitive when younger or older, possibly because of a lower rate of proliferation of target cells.

3.3.2 Role of Immune System

While a comprehensive review of this area of research is beyond the scope of this Report, there has been no overwhelming evidence thus far to suggest that immunity plays a consistent or decisive role in protecting against experimental chemical carcinogenesis; on the contrary, a number of studies have suggested that, under certain circumstances, an intact immune system may actually enhance the process of carcinogenesis in experimental animals (Prehn, 1977). Nevertheless the immune system has been studied extensively as a major endogenous modifying factor (Melief and Schwartz, 1975), and there is ample evidence that immunosuppression may enhance susceptibility to certain cancers in animals and man (Herberman, 1984).

3.3.3 Influence of Genetic Background

Susceptibility to experimental chemical carcinogenesis varies according to genetic background, sex, species, and strain. Animal species differ in organ site specificity for particular carcinogens, although most carcinogens induce tumors in more than one species, and some, such as N-nitroso compounds, induce tumors in every species. Strain specific differences also exist within a given species (Kouri and Nebert, 1977), and animals have been bred for sensitivity at a particular organ site (Baird and Boutwell, 1974). Such species and strain differences in susceptibility can be explained in some cases by the genetic control of metabolic capabilities for activating chemical carcinogens. Differences in susceptibility for nonendocrine related organ sites also are sexually determined in animal models. Liver cancer induced by 2-acetylaminofluorene (2-AAF), aflatoxin B_1, and azo dyes is more prevalent in male rats because of androgen-modulated enzymatic activity (Weisburger and Williams, 1975). However, metabolic differences are not the principal determinant of susceptibility in all cases, and little is known of the cellular or molecular nature of genetic sensitivities.

3.3.4 Anticarcinogens

Anti-carcinogens decrease the risk of cancer by diminishing the absolute tumor incidence or lengthening the tumor latency period. There are many examples of anticarcinogens in experimental studies (Wattenberg, 1978). Anticarcinogens may exert their effects at all stages of neoplastic development. Since most chemical carcinogens require metabolic activation, usually via oxidative pathways, antioxidants such as ascorbate and butylated hydroxytoluene may prevent neoplastic transformation by trapping the ultimate carcinogenic species and/or by inhibiting the formation of the ultimate carcinogens.

Agents such as glutathione may act as scavengers for electrophilic ultimate carcinogens. During neoplastic progression, anticarcinogens may act directly on the transformed cell, altering its biological potential and/or enhancing the host defense system against the cell. This group of agents includes hormones and retinoids (analogs of vitamin A), (Sporn, 1977).

3.4 Dose-Response Relationships

A wide diversity of dose-response (incidence) relationships has been observed among the neoplasms induced experimentally by chemicals (Zeise *et al.*, 1987), radiation (UNSCEAR, 1977) or both. Although neoplasms of virtually every type have been induced in one experiment or another, not all types of neoplasms are observed in animals of any one species or strain. Under some conditions, moreover, the incidence of certain neoplasms has actually been observed to decrease with increasing dose of whole-body irradiation (see Figure 3.1).

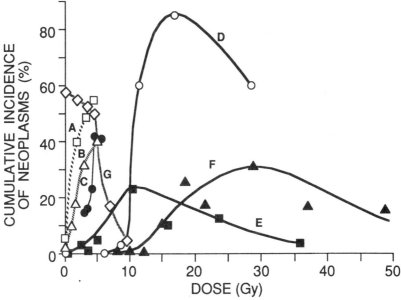

Fig. 3.1 Dose-incidence curves for different neoplasms in animals exposed to external radiation: (A) myeloid leukemia in x-irradiated mice (□) (Upton *et al.*, 1958); (B) mammary gland tumors at 12 months in gamma-irradiated rats (△) (Shellabarger *et al.*, 1969); (C) thymic lymphoma in x-irradiated mice (●) (Kaplan and Brown, 1952); (D) kidney tumors in x-irradiated rats (o) (Maldague, 1969); (E) skin tumors in alpha-irradiated rats (percentage incidence x 10) (■) (Burns *et al.*, 1968); (F) skin tumors in electron-irradiated rats (percentage incidence x 10) (▲) (Burns *et al.*, 1968); (G) reticulum cell sarcoma in x-irradiated mice (◊) (Metalli *et al.*, 1974); (H) lung adenomas in neutron-irradiated mice (*) (Ullrich *et al.*, 1976) (Modified from UNSCEAR, 1972) (from Upton, 1984).

14 / 3. CARCINOGENIC EFFECTS OF RADIATION AND CHEMICALS

Some of the observed variations among chemically-induced neoplasms (e.g., Figure 3.2) are attributable at least in part to species-, strain-, organ-, and sex-dependent differences in pharmacokinetics which influence the dosage of carcinogen at the biological target. Such

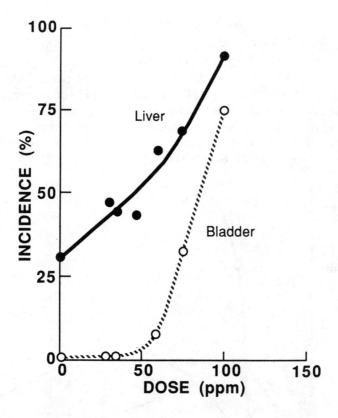

Fig. 3.2 Cumulative incidence of neoplasms of the liver and urinary bladder in female BALB/c mice exposed to 2 acetylamino fluorene at various concentrations in the diet for up to 33 months (from Littlefield et al., 1979).

an explanation cannot account, however, for the observed variations in dose-incidence relations among radiation-induced neoplasms (see Figure 3.1), which remain to be explained.

3.4 DOSE-RESPONSE RELATIONSHIPS / 15

The diversity of observed dose-incidence relationships is not astonishing in view of the multicausal, multistage nature of carcinogenesis. Although the different dose-incidence curves cannot be represented by a single mathematical function, they exhibit a number of features in common.

First, the carcinogenic effects of radiation and chemicals are expressed as an increase in the age-specific incidence of particular neoplasms. Depending on the survival of the population at risk, the final cumulative incidence of neoplasms in the population may, or may not, be increased.

Second, for the induction of a given neoplasm, the carcinogenic effectiveness of chemicals varies in relation to molecular structure, and the carcinogenic effectiveness of radiation varies in relation to LET. In laboratory animals, carcinogenic potency is measured in terms of the total number of tumors produced per group or per animal, percentage of animals developing tumors, period of latency (time between exposure and appearance of tumor), or some combination of these measurements. For example, Iball's carcinogenesis index (C.I.) represents the percentage of animals with tumors following the injection of a fixed dose, divided by the latency period in days, multiplied by 100 (Iball, 1939). The Salmonella/liver microsome mutagenesis assay (Ames *et al.*, 1975) attempts to quantify the mutagenic effects of various chemical compounds (or various potential carcinogens) on the basis of the number of nanomoles of a test compound required to double the reversion frequency over background rates in selected strains of Salmonella. Numerous other tests attempting to quantify mutagenic and carcinogenic potentials of various chemicals are detailed in a National Academy of Sciences/National Research Council report (NAS/NRC, 1983a). On the basis of the daily dose required to halve the probability of remaining free of tumors throughout life in rats or mice, chemicals have been observed to vary in potency by more than six orders of magnitude (Gold *et al.*, 1984). With radiation, the dose-incidence curve generally rises more steeply with dose in the case of high-LET radiation than in the case of low-LET radiation, especially at low dose rates (see Figure 3.3). The dose incidence curves for radiation and chemicals also vary in shape as well as slope, depending on the type of neoplasm induced and the carcinogenic chemical in question, e.g., Figures 3.1. and 3.2.

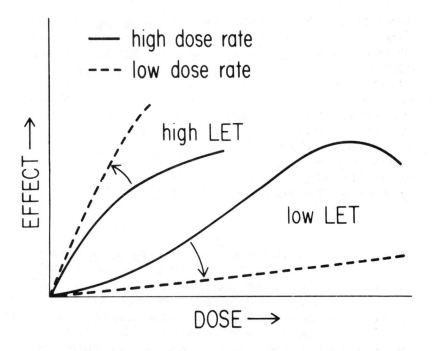

Fig 3.3. Schematic dose-response curves for incidence of tumors in relation to dose and dose rate of high-LET and low-LET radiation. (from Upton, 1984; also see Thomson et al., 1982; Sinclair, 1983).

Third, for many types of neoplasms, the incidence passes through a maximum at some intermediate dose and decreases as the dose is increased further, at least when the exposure is accumulated at a high dose rate; e.g., Figure 3.1.

Fourth, with chronic, daily exposures, the *median time (t)* of tumor appearance tends to vary inversely with *daily dose (d)*, within limits, according to the function:

$$dt^n = \text{constant} \qquad (3.1)$$

where n is always greater than one (Albert and Altshuler, 1973). Furthermore, in contrast to the situation with low-LET radiation, where prolongation of exposure characteristically reduces the carcinogenic effectiveness of a given total dose (Figure 3.3), the carcinogenic effectiveness of a given dose of a complete chemical carcinogen tends to depend less, if at all, on the duration of administration, or exposure.

Fifth, because of the different types of effects through which radiation and cytotoxic chemicals can influence carcinogenesis at high dose levels, the dose-incidence curves may reflect various combinations of initiating effects, promoting effects, and anticarcinogenic effects, depending on the dose, dose rate, and agent in question, e.g., Figure 3.4.

Sixth, interactive effects of radiation and/or chemical carcinogens with other physical and chemical agents may be additive, synergistic, or antagonistic, depending on the agents in question and the conditions of exposure (UNSCEAR, 1982). Because of the complexity of these interactions, which have received relatively little study to date, it is not possible to predict in advance the combined effects of two or more agents on the basis of theory or experience with other combinations of agents (UNSCEAR, 1982).

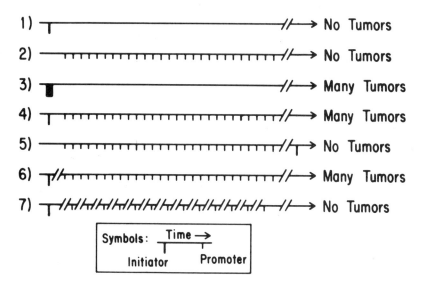

Fig. 3.4 Schematic representation of the tumor-inducing effects of exposure to an initiating agent, a promoting agent, or both (from Boutwell, 1978).

4. Nature, Distribution, and Sources of Ionizing Radiation in the Environment

4.1 Physical Properties of Ionizing Radiation

Ionizing radiations include electromagnetic waves (photons) of high frequency, and particles of varying mass and charge, e.g., electrons, protons, neutrons, and alpha particles, travelling at initial velocities approaching the speed of light. In contrast to other forms of radiation, each ionizing radiation track is able to deposit enough localized energy in an absorbing material to disrupt atoms and molecules in its path. The resulting atomic changes, e.g., the ejection of an electron from its orbit around the nucleus of an atom, or the ejection of a proton from the nucleus of an atom, can produce ion pairs. The ion pairs, in turn, may lead to molecular damage and, ultimately, biological injury.

4.2 Sources and Levels in the External and Internal Environment

Life on earth has evolved in the presence of naturally occurring ionizing radiation, which is continuous and ubiquitous. In addition to natural background radiation exposure, mankind is now exposed also to radiation from various man-made sources.

Natural background radiation consists mainly of: (1) Radon and radon daughters, (2) cosmic rays, (3) cosmogenic radiation (4) terrestrial radiation, and (5) internal radiation (potassium-40, carbon-14, and other naturally-occurring radionuclides in the body). The average dose rate from these sources received by a person living at sea level is about 3 mSv (300 mrem) effective dose equivalent per year (see Table 4.1). The dose may be larger at higher altitudes because of increased cosmic ray intensities, and it may be larger in areas where the content of natural radioactive material in the soil is increased.

Exposure to radiation from man-made sources is estimated to deliver an average annual effective dose equivalent of about 0.6 mSv (60 mrem) to the general population (Table 4.1). The largest contribution comes

TABLE 4.1—*Estimates of annual effective dose equivalent to members of the U.S. population*[a]

Source	Average annual H_E[b] mSv y⁻¹ (mrem y⁻¹)	
Natural sources		
Radon and radon daughters	2.0	(200)
Cosmic rays	0.27	(27)
Cosmogenic radiation	0.01	(1)
Terrestrial radiation	0.28	(28)
Internal radiation	0.40	(40)
Man-made		
Nuclear fuel cycle	0.0005	(0.05)
Miscellaneous environmental	0.0006	(0.06)
Medical		
Diagnostic x rays	0.39	(39)
Nuclear medicine	0.14	(14)
Occupational	0.009	(0.9)
Consumer products	0.05-0.13	(5-13)
Rounded Total	3.60	(360)

[a] From NCRP, 1987a
[b] H_E is the effective dose equivalent

from medical diagnosis. Lesser contributions come from "technologically enhanced" sources, e.g., phosphate fertilizers and building materials containing trace amounts of natural radioactivity, consumer products, e.g., color TV sets, smoke detectors, luminescent clock and instrument dials.

In comparison with the average effective dose equivalent to the general population from cosmic radiation, which approximates 0.27 mSv (27 mrem) annually, the average dose to the thorax of an individual from a standard x-ray examination of the chest is of the order of 0.10 mSv (10 mrem). Other diagnostic procedures may deliver substantially larger doses, e.g., the average effective dose equivalent of an individual from a barium enema examination of the colon is approximately 4 mSv (400 mrem).

4.3 Assessment of Exposure to Ionizing Radiation

4.3.1 *Introduction*

Units used for radiation exposure, radiation dose and radiation dose equivalent are in transition (ICRU, 1980; NCRP, 1985). The relationship between the SI units and the conventional units are given in Table 4.2.

To characterize the distribution of a dose in time, the terms *dose rate* and *dose fractionation* are used. Dose rate is customarily expressed in Gy(rad) per minute, or per hour, or in comparable units. The term

4. NATURE, DISTRIBUTION AND SOURCES OF IONIZING RADIATION

TABLE 4.2—*Units*[a]

Quantity	SI Unit	Conventional unit, symbol & value in SI Units
Exposure	no special name (C kg^{-1})	1 roentgen (R) = 2.58 x 10^{-4} C kg^{-1}
Absorbed Dose	1 gray (Gy) = 1 J kg^{-1}	1 rad (rad) = 0.01 Gy
Dose Equivalent	1 sievert (Sv)[b] = 1 J kg^{-1}	1 rem (rem)[b] = 0.01 Sv

[a] Where the abbreviations C, kg and J in the table stand for coulomb, kilogram and joule respectively.
[b] The magnitude of the dose equivalent in sievert or rem depends on the LET of the radiation (see text).

fractionation denotes the division of a dose into two or more successive increments separated from one another in time.

The term *linear energy transfer* (LET) (ICRU, 1980) is used to characterize the distribution of ionizing events along the path of impinging radiation. These events tend to be sparsely distributed along the track of a photon as it traverses a cell, whereas they are densely distributed along the track of an alpha particle. Consequently, x rays and gamma rays are characteristically referred to as low-LET radiations, while alpha particles and other densely ionizing particles are referred to as high-LET radiations.

The relative biological effectiveness (RBE) of a radiation generally increases with LET, to an extent that depends on the biological effect in question and the conditions of irradiation (ICRU, 1970). For purposes of radiological protection, however, a single *quality factor (Q)* is usually used for converting the absorbed dose into dose equivalent (ICRP, 1977; NCRP, 1987b). The values of Q for radiations of different LETs range from 1 for 250 kVp x rays or gamma rays to 20 for alpha particles (NCRP, 1987b).

The unit used for expressing the collective dose to a population is the person-sievert (person-rem), which is the product of the number of people exposed times the average dose per person, e.g., 10 mSv to each of 1,000 people = 10,000 person-mSv = 10 person-Sv (1 rem to each of 1,000 people = 1,000 person-rem).

In expressing the dose to a tissue, it is customary to average the dose over the entire tissue, although in practice the dose is seldom uniform over an entire tissue or organ. Strictly speaking, of course, the averaging of a non-uniform dose over a tissue or organ cannot be expected to be valid for calculation of dose-effect relationships unless the relationships in question are linear and the number of cells at risk vary in proportion to the fraction of tissue exposed. With low doses, only a small percentage of the cells in a tissue, or of the relevant structures within the cell, are hit at all, and this number increases in proportion with the dose.

The conventional unit of radioactivity, the curie (Ci), is equivalent to 3.7×10^{10} radioactive events per second. The SI unit used for denoting the amount of radioactive material contained in a given sample of matter is the becquerel (Bq); one becquerel is that quantity of a radioactive nuclide in which there is one radioactive event per second (1 Bq = 2.7×10^{-11} Ci). Since radionuclides decay exponentially with time, each element at its own rate, the time required for a given quantity of a radionuclide to lose one-half of its radioactivity is call its *physical half-life*.

The dose of radiation delivered by an internally deposited radionuclide depends on the quantity of radioactive material residing *in situ*. This quantity decreases as a function of the physical half-life of the radionuclide and the rate at which the element is redistributed or excreted (i.e., its biological half-life). Because the physical half-life is known precisely and the biological half-life can be characterized within limits for most radionuclides, the dose to a tissue that will ultimately be delivered by a given concentration of a radionuclide deposited therein can be predicted to a first approximation. The collective dose to a population that will be delivered by the radionuclide—the so-called collective dose commitment—serves as the basis for assessing the relevant long-term health effects of the nuclide.

4.3.2 *Route of Exposure and Distribution of Absorbed Dose*

Energetic x rays and gamma rays penetrate tissue well enough so that an external source of such radiations may deliver a relatively uniform dose to all parts of the body. Particulate radiations, on the other hand, penetrate much less readily, especially those with high mass and charge, such as alpha particles. With such radiations, irradiation of deep-seated tissues will occur only as a result of internal deposition.

One of the most important factors influencing the uptake and distribution of radioactive substances is the route of exposure. For essentially all radionuclides, both rate of uptake and organ distribution may vary, depending on whether exposure results from entry via the gastrointestinal, respiratory, or cutaneous route. Additional factors influencing the uptake and/or clearance of radionuclides include their physicochemical state, particle size, and solubility, as well as the presence of carrier substances or other co-factors. Insoluble substances tend to be localized in phagocytes, forming "hot spots" where the concentration of radioactive material may be orders of magnitude higher than that in surrounding tissues. Even in the case of soluble radionuclides, localization within cells and tissues may be highly nonuniform, depending on metabolic factors.

4.3.3 *Metabolism: Influence on Distribution of Radionuclides in the Body*

Metabolically, radionuclides are handled in the same way as stable elements of the same atomic number. Thus radioactive iodine simulates stable iodine, being concentrated in the thyroid gland so predictably that its rate of uptake provides an accurate measure of thyroid function. The metabolism of other radionuclides also is sufficiently characteristic so that their patterns of uptake, distribution, translocation, and excretion are similarly predictable.

Although the rates at which radionuclides of different elements are eliminated from an organ or from the body vary widely, all patterns can be represented to a first approximation by a series of exponential functions. From these functions the characteristic "biological half-time" of each radionuclide can be derived. For iodine-131 the *biological half-life* in the thyroid gland is about 90 days in adults. Hence, since the *physical half-life* is about eight days, the *effective half-life* (i.e., the time in which the amount of radioactive material remaining *in situ* decreases by one-half) is between seven and eight days. For most bone-seeking radionuclides, the biological half-times are appreciably longer. In the case of plutonium-239, for example, the biological half-life is 50 years or more, and the physical half-life is some 24,000 years.

While the effects of internally deposited radionuclides are attributable in most instances to the type and energy of the radiations they emit, their transmutation on decay has also been implicated in certain circumstances, e.g., the severe mutagenicity and lethality caused by decay of iodine-125 in 5-iododeoxyuridine-labelled DNA is attributed in part to rupture of the uracil ring following transmutation of the iodine (Stocklin, 1979).

4.4 Summary

All living forms are exposed to natural background radiation from extra-terrestrial, terrestrial, and internal sources. The effective dose equivalent from these sources to humans, which varies substantially with geographic location, averages about 3 mSv (300 mrem) per year. Approximately 2 mSv (200 mrem) of the 3 mSv (300 mrem) due to natural background radiation is due to the inhalation of radon and its daughters. In addition to natural background radiation, populations are exposed to radiation as a result of underground mining, medical, dental, and industrial radiography, phosphate fertilizers, radioactive weapons fallout, and nuclear power production. Collectively, these and other man-made sources of exposure are estimated to deliver an average effective dose equivalent of about 0.6 mSv (60 mrem) per year to members of the general population. The average tissue doses from most sources are comparatively well characterized for the population at large. The doses from certain internally deposited radionuclides, however, vary with space, time, route of exposure, metabolism, diet, rate of growth, and other variables. Hence, these variables must be taken into account when applicable.

5. Nature, Distribution, and Sources of Carcinogenic Chemicals in the Environment

5.1 Chemical Properties

It is not possible on the basis of present knowledge of the relationship of chemical structure to the carcinogenic activity of a chemical to predict how many of the millions of compounds in nature, or the tens of thousands of compounds in commerce, are carcinogenic. Although relatively few chemicals have been observed to cause cancer in human populations, those that have (Table 5.1), and the hundreds of other substances for which there is some evidence of carcinogenicity in laboratory animals (Table 5.2), include compounds of widely diverse structures.

Although our knowledge of the relationship between molecular structure and carcinogenic activity is limited, known carcinogens can be classified into subcategories on the basis of other criteria. Thus, "direct-acting" carcinogens, which are active in the form in which they are encountered in the environment, can be distinguished from "indirect-acting" carcinogens, which require metabolic conversion into their active derivatives, a process that is under genetic control. Thus liver cancer induced by AAF, aflatoxin B_1, and azo dyes is more prevalent in male rats because of androgen-modulated enzymatic activity (Weisburger and Williams, 1975). Metabolic differences are not the principal determinant of susceptibility in all cases, however, and little is known of the cellular or molecular nature of genetic sensitivities.

5.2 Sources and Levels of Carcinogenic Chemicals in the External Environment

A brief examination of Tables 5.1 and 5.2 shows many chemicals that are common in the environment. In contrast to the relatively extensive assessment of human exposure to ionizing radiation, assessment of the extent of human exposures to these and other chemicals is fragmentary. Because there are some 6,000,000 known chemicals (NAS/NRC,

5.2 SOURCES AND LEVELS OF CARCINOGENIC CHEMICALS / 25

1984)—60,000 in commerce—the task of measurement is huge. Also, whereas radioactive materials decay with well defined half-lives, the fates of chemicals are much more diverse.

Exposure estimates for chemicals have been limited to relatively few cases, e.g., (1) calculations of exposures following dispersion of known amounts of a gaseous chemical from a chimney stack, (2) measurements of dominant pollutants—such as sulfates—in polluted air, and (3) measurements of concentrations in water, particularly drinking water. There is concern for, but very little measurement of, the concentrations of chemicals leaching from toxic waste dumps. Even in these limited cases, the measurements are restricted for practical reasons largely to pollutants present in relatively high concentrations (greater than $1\mu g$ per liter for drinking water) or to potent carcinogens that are believed to be present. It is probable that the measurements may miss carcinogens of moderate carcinogenic potency that are present at low levels only.

5. NATURE, DISTRIBUTION AND SOURCES OF CARCINOGENIC CHEMICALS

TABLE 5.1—*Chemicals and related industrial processess for which there are evidence for carcinogenicity in humans*[a]

Group 1[b]	
Alcohol (ethanol)	Diethylstilbestrol
4-Aminobiphenyl	
Arsenic; certain arsenic compounds	
	Isopropyl oil (mfgr)
Asbestos	(Manufacture of isopropyl alcohol by the strong acid process)
Benzene	Mustard gas
Benzidine	2-Naphthylamine
N, N-bis (2-chloroethyl-2-naphthyl amine (chlornaphazine)	Nickel refining
Bis (chloromethyl) ether; technical grade chloromethyl methyl ether	Soots, tars, and mineral oils
Chlormaphazine	Vinyl chloride
Chromium; certain chromium compounds	
Cigarette smoke	
Group 2[c]	
Subgroup A[d]	
Aflatoxins	Cyclophosphamide
Cadmium, certain cadmium compounds	Nickel; certain nickel compounds
Chlorambucil	Tris (1-aziridnyl) phosphine sulphide (thiotepa)
Subgroup B	
Acrylonitrile	Dimethylsulphate
Amitrole (Aminotriazole)	Ethylene oxide
Auramine	Iron dextran
Beryllium; certain beryllium compounds	Oxymetholone
Carbon tetrachloride	Phenacetin
Dimethylcarbamoyl chloride	Polychlorinated biphenyls

[a] From IARC, 1982; also Schottenfeld and Haas, 1978; Doll and Peto, 1981.

[b] For the chemicals or industrial processes in Group 1, there appears to be convincing epidemiologic evidence of carcinogenicity in humans.

[c] Evidence for the chemicals in Group 2 indicates that they are *probably* carcinogenic in humans.

[d] Evidence for chemicals in subgroup A is stronger than that for chemicals in subgroup B.

5.2 SOURCES AND LEVELS OF CARCINOGENIC CHEMICALS / 27

TABLE 5.2—*Partial list of chemicals in IARC monographs 1-16, including the 221 for which there was some evidence of carcinogenicity in laboratory animals only*[a]

Note: asterisk (*) denotes the 221 chemicals for which there was some evidence of carcinogenicity in laboratory animals only. For the remaining 121 chemicals, the available data were inadequate to enable an evaluation of carcinogenicity in animals or humans. Excluded from the list are the chemicals for which there was sufficient evidence of carcinogenicity in humans (see Table 5.1).

1.	Acetamide*	40.	BHC (technical grades)*
2.	Acridine orange	41.	Bis (1-aziridinyl)-morpholino-phosphine sulfide*
3.	Acriflavinium chloride		
4.	Actimomycins*	42.	Bis (chloroethyl) ether*
5.	Adriamycin	43.	1, 2-Bis (chloromethoxy) ethane*
6.	Aldrin	44.	1, 4-Bis (chloromethoxymethyl)-benzene*
7.	Amaranth		
8.	5-Aminoacenaphthene	45.	Blue VRS*
9.	p-Aminoazobenzene*	46.	Brilliant blue FCF*
10.	o-Aminoazotoluene*	47.	1, 4-Butanediol dimethane-sulfonate (Myleran)*
11.	p-Aminobenzoic acid		
12.	2-Amino-5-(5-nitro-2-furyl)-1,3,4-thiadiazole*	48.	b-Butyrolactone*
		49.	γ-Butyrolactone
13.	4-Amino-2-nitro-phenol	50.	Cadmium acetate
14.	Amitrole*	51.	Cadmium chloride*
15.	Aniline	52.	Cadmium powder*
16.	Anthranilic acid	53.	Cadmium sulfate*
17.	Apholate	54.	Cadmium sulfide*
18.	Aramite*	55.	Calcium arsenate
19.	Arsenic Trioxide	56.	Calcium chromate*
20.	Aurothioglucose*	57.	Cantharidin*
21.	Azaserine*	58.	Carbaryl
22.	Axiridine*	59.	Carbon tetrachloride*
23.	2-(1-Aziridinyl)-ethanol*	60.	Carmoisine
24.	Aziridyl benzoquinone*	61.	Catechol
25.	Azobenzene*	62.	Chlorambucil*
26.	Barium chromate	63.	Chlorinated dibenzodioxins
27.	Benz[a]acridine*	64.	Chlormadinone acetate*
28.	Benz[c]acridine*	65.	Chlorobenzilate*
29.	Benzo[b]fluoranthene*	66.	Chloroform
30.	Benzo[j]fluoranthene*	67.	Chloropropham
31.	Benzo[a]pyrene*	68.	Chloroquine
32.	Benzo[e]pyrene*	69.	p-Chloro-o-toluidine (hydrochloride)
33.	Benzyl chloride*		
34.	Benzyl violet 4B*	70.	Cholesterol
35.	Beryllium*	71.	Chromic chromate*
36.	Beryllium oxide*	72.	Chromium acetate
37.	Beryllium phosphate*	73.	Chrysene*
38.	Beryllium sulfate*	74.	Chrysoidine*
39.	Beryl ore*	75.	C.I. Disperse Yellow 3

[a]From IARC Monographs, Volumes 1 to 16; Tomatis, *et al.*, 1978

5. NATURE, DISTRIBUTION AND SOURCES OF CARCINOGENIC CHEMICALS

76. Cinnamyl anthranilate
77. Citrus Red No. 2*
78. Copper 8-hydroxyquinoline
79. Coumarin*
80. Cycasin*
81. Cyclochlorotine*
82. 2, 4-D and esters
83. Daunomycin*
84. D & C Red No. 9
85. Dichlorodiphenyldichloroethane (DDD)
86. 1, 1-Dichloro-2-2-bis (p-chlorophenyl) ethylene (DDE)
87. DDT*
88. Diacetylaminoazotoluene
89. N, N-Diacetylbenzidine*
90. Diallate*
91. 2, 4-Diaminoanisole (sulfate)
92. 4, 4-'-Diaminodiphenyl ether*
93. 1, 2-Diamino-4-nitrobenzene
94. 1, 4-Diamino-2-nitrobenzene
95. 2, 6-Diamino-3-(phenylazo)-pyridine (hydrochloride)
96. 2, 4-Diaminotoluene*
97. 2, 5-Diaminotoluene (sulfate)
98. Diazepam
99. Diazomethane*
100. Dibenz(a,h) acridine*
101. Dibenz(a,j) acridine*
102. Dibenz(a,h) anthracene*
103. Dibenzo(c,g) carbazone*
104. Dibenzo(h,rst) pentaphene*
105. Dibenzo(a,e)pyrene*
106. Dibenzo(a,h)pyrene*
107. Dibenzo(a,i)pyrene*
108. Dibenzo(a,l)pyrene*
109. 1, 2-Dibromo-3-chloropropane*
110. Dibutylnitrosamine*
111. o-Dichlorobenzene
112. p-Dichlorobenzene
113. 3, 3'-Dichlorobenzidine*
114. trans-Dichlorobutene
115. 3, 3'-Dichloro-4, 4'-diaminodiphenyl ether*
116. Dieldrin*
117. Diepoxybutane*
118. 1, 2-Diethylhydrazine
119. Diethylnitrosamine*
120. Diethyl sulfate*
121. Diglycidyl resorcinol ether
122. Dihydrosafrole*
123. Dimethisterone
124. Dimethoxane*
125. 3, 3'-Dimethoxybenzidine*
126. p-Dimethylaminoazobenzene*
127. p-Dimethylaminobenzenediazosodium sulfonate
128. trans-2-[Dimethylamino) methylamino]-5-[2-5-nitro-2-furyl) vinyl]-1,3,4-oxadiazole*
129. 3, 3'-Dimethylbenzidine*
130. Dimethylcarbamoyl chloride*
131. 1, 1-Dimethylhydrazine*
132. 1, 2-Dimethylhydrazine*
133. Dimethylnitrosazine*
134. Dimethyl sulfate*
135. Dinitrosopentamethylene-tetramine
136. 1, 4-Dioxane*
137. 2, 4'-Diphenyldiamine
138. Disulfiram
139. Dithranol*
140. Dulcin
141. Endrin
142. Eosin (disodium salt)
143. Epichlorohydrin*
144. 1-Epoxyethyl-3, 4-Epoxycyclohexane*
145. 3, 4-Epoxy-6-methylcyclohexylmethyl-3, 4-epoxy-6-methyl carboxylate*
146. cis-9, 10-epoxy-stearic acid
147. Estradiol mustard*
148. Ethinylestradiol*
149. Ethionamide*
150. Ethylene dibromide*
151. Ethylene oxide
152. Ethylene sulfide*
153. Ethylenethiourea*
154. Ethyl methane-sulfonate*
155. Ethyl Selenac
156. Ethyl Tellurac
157. Ethynodiol diacetate*
158. Evans blue*
159. Fast green FCF*
160. Ferbam
161. 2-(2-Formylhydrazino)-4-(5-nitro-2-furyl) thiazole*

5.2 SOURCES AND LEVELS OF CARCINOGENIC CHEMICALS / 29

162. Fusarenon-X
163. Glycidaldehyde*
164. Glycidyl oleate
165. Glycidyl stearate
166. Griseofulvin*
167. Guinea green B*
168. Heptachlor
169. Hexamethylphosphoramide*
170. Hycathone (mesylate)*
171. Hydrazine*
172. Hydroquinone
173. 4-Hydroxyazobenzene
174. 8-Hydroxyquinoline
175. Hydroxysenkikine
176. Indeno(1,2,3-cd)pyrene*
177. Iron dextran*
178. Iron dextrin*
179. Iron oxide
180. Iron-sorbitol-citric acid complex
181. Isatidine*
182. Isonicotinic acid hydrazide*
183. Isopropyl alcohol
184. Isosafrole*
185. Jacobine
186. Lasiocarpine*
187. Lead acetate*
188. Lead arsenate
189. Lead carbonate
190. Lead chromate
191. Lead phosphate*
192. Lead subacetate*
193. Ledate
194. Light green SF*
195. Lindane*
196. Luteoskyrin*
197. Magenta*
198. Maleic hydrazine*
199. Maneb
200. Mannomustine (dihydrochloride)*
201. Medphalan
202. Medroxyprogesterone acetate*
203. Merphalan*
204. Mestranol*
205. Methoxychlor
206. 2-Methylaziridine*
207. Methylazoxymethanol acetate*
208. Methyl carbamate
209. N-Methyl-N, 4-dinitrosoaniline*
210. 4, 4'-Methylenebis-(2-chloroaniline)*
211. 4, 4'-Methylenebis-(2-methylaniline)*
212. 4, 4'-Methylenedianiline
213. Methyl iodide*
214. Methyl methane-sulfonate*
215. N-Methyl-N'-nitro-N-nitrosoguanidine*
216. Methyl red
217. Methyl Selenac
218. Methylthiouracil*
219. Metronidazole*
220. Mirex*
221. Mitomycin C*
222. Monocrotaline*
223. Monuron*
224. 5-(Morpholino-methyl)-3-[(5-nitrofurfurylidene)-amino]-2-oxazolidinone*
225. 1-Naphthylamine*
226. Native carrageenans*
227. Nickel carbonyl*
228. Nickelocene*
229. Nickel oxide*
230. Nickel powder
231. Nickel subsulfide*
232. Niridazole*
233. 5-Nitroacenaphthene*
234. 4-Nitrobiphenyl*
235. Nitrofuraldehyde semicarbazone
236. 1-[(5-Nitrofurfurylidene) amino]-2-imidazolidinone*
237. N-[(4-5-Nitro-2-furyl)-2-thiazolyl]-acetamide*
238. Nitrogen mustard (hydrochloride)*
239. Nitrogen mustard N-oxide (hydrochloride)*
240. Nitrosoethylurea*
241. Nitrosomethylurea*
242. N-Nitroso-N-methylurethan*
243. Norethisterone*
244. Norethisterone acetate*
245. Norethynodrel*
246. Norgestrel

5. NATURE, DISTRIBUTION AND SOURCES OF CARCINOGENIC CHEMICALS

247. Ochratoxin A
248. 17 β-Oestradiol*
249. Oestriol
250. Oestrone*
251. Oil orange SS*
252. Orange 1*
253. Orange G
254. Oxazepam*
255. Oxyphenbutazone
256. Parasorbic acid*
257. Patulin*
258. Penicillic acid*
259. Phenicarbazide*
260. Phenobarbital sodium*
261. Phenoxybenzamine*
262. Phenylbutazone
263. m-Phenylenediamine (hydrochloride)
264. p-Phenylenediamine (hydrochloride)
265. N-Phenyl-2-naphthylamine*
266. Polychlorinated biphenyls*
267. Ponceau MX*
268. Ponceau 3R*
269. Ponceau SX
270. Potassium arsenite
271. Potassium bis (2-hydroxyethyl)-dithiocarbamate*
272. Progesterone*
273. Pronetalol hydrochloride*
274. 1, 3-Propanesultone*
275. Propham
276. β-Propiolactone*
277. n-Propyl carbamate*
278. Propylene oxide*
279. Propylthiouracil*
280. Pyrimethamine*
281. p-Quinone
282. Quintozene*
283. Reserpine
284. Resorcinol
285. Retrosine*
286. Rhodamine B*
287. Rhodamine 6G*
288. Riddelline
289. Saccharated iron*
290. Safrole*
291. Scarlet red
292. Selenium compounds
293. Semicarbazine (hydrochloride)*
294. Seneciphylline
295. Senkirkine
296. Sodium arsenate
297. Sodium arsenite
298. Sodium dichromate
299. Sodium diethyldithio-carbamate
300. Sterigmatocystin*
301. Streptozotocin*
302. Strontium chromate*
303. Styrene oxide
304. Succinic anhydride*
305. Sudan I*
306. Sudan II*
307. Sudan III
308. Sudan brown RR
309. Sudan red 7B
310. Sunset yellow FCF
311. 2,4,5-T and esters
312. Tannic acid*
313. Terpene polychlorinates*
314. Testosterone*
315. Tetraethyl & tetramethyl lead
316. Thioacetamide*
317. 4,4-Thioaniline*
318. Tiouracil*
319. Tiourea*
320. Thiram
321. o-Toluidine (hydrochloride)
322. Trichloroethylene*
323. Trichlorotriethylamine hydrochloride
324. Triethylene glycol diglycidyl ether*
325. Tris(aziridinyl)-p-benzoquinone*
326. Tris(1-aziridinyl)-phosphine oxide
327. Tris(1-aziridinyl)-phosphine sulfide*
328. 2,4,6-Tris(1-aziridinyl)-s-triazine*
329. 1,2,3-Tris(chloromethoxy)-propane*
330. Tris(2-methyl-1-aziridinyl) phosphine oxide
331. Trypan blud*
332. Uracil mustard*
333. Urethan*
334. Vinyl cyclohexane

335.	2,4-Xylidine (hydrochloride)	339.	Zectran
336.	2,5-Xylidine (hydrochloride)	340.	Zinc chromate hydroxide*
337.	Yellow AB	341.	Zineb
338.	Yellow OB*	342.	Ziram

5.3 Assessment of Exposure

5.3.1 *Units of Measurement, Dosimetry*

Chemicals can be quantified by weight, e.g., grams, or by number of molecules, e.g., mols. One mol, which is equivalent to the molecular weight in grams of the chemical being considered, represents about 6×10^{23} molecules. Fractions of a mol are commonly expressed as: millimol (10^{-3} mol), micromol (10^{-6}), nanomol (10^{-9}), picomol (10^{-12}), and femtomol (10^{-15} mol). Hence, one femtomol (fmol) represents about 6×10^{8} (or 600 million) molecules.

As is the case with irradiation, specification of exposure to chemical carcinogens must include consideration of the size of the animal species, maturity of the organism, and the duration of exposure. Clinical patients or laboratory animals are administered drugs or other chemicals on a milligram-per-kilogram of body weight basis, or a milligram-per-square-meter basis, at specified time intervals. The intensity and duration of the responses are dependent on both the size of the dose and the frequency with which the chemical is administered. The dose and frequency are of greater importance and more variable with chemicals than with radiation, due to: (1) large differences among individuals in *ability to metabolize* chemicals and (2) the phenomenon of enzyme *induction* (repeated doses of sufficient size can cause enzyme levels to increase, thereby allowing more rapid metabolism of the subsequent doses of the chemical and other substrates of related molecular structure) (Nebert and Gonzalez, 1985). Thus, the half-life of the chemical (or its reactive intermediates) is dependent upon its rate of metabolism in each tissue, which can vary greatly from one tissue to another and across time.

5.3.2 *Ambient Levels in the Environment*

Various polycyclic aromatic hydrocarbons, carbonyl arenes, polycyclic nitro-aromatics, aza arenes, N-nitroso compounds, and other carcinogens have been detected in ambient air (Hughes *et al.*, 1980; Kneip *et al.*, 1983; Spiegelhalder and Preussman, 1983). Diesel exhaust particles

(Clark and Vigil, 1980) are known to be extremely mutagenic in the *Salmonella*/liver microsomes *in vitro* assay. Benzo[a]pyrene, for example, has been monitored the most extensively and is frequently used as an indicator of polycyclic aromatic hydrocarbon contamination; however, the composition of polycyclic organic matter from different sources can vary widely. Annual average ambient benzo[a]pyrene concentrations in urban areas around the U.S. during the past 20 years range from less than 0.1 to more than 30 ng m^{-3}. Regions near petroleum and coal-burning installations, distilleries, and automobile exhausts are usually among those with high readings. The highest recording, 120 ng per cubic meter, was in a smoke-filled room in Prague, demonstrating the relative large quantities of this contaminant in cigarette smoke. Air pollution generally is greatest in the winter, presumably due to heating sources and the increased likelihood of atmospheric inversion layers. The trend during the past two decades suggests that concentrations of pollutants in industrial and urban areas are decreasing, perhaps due to increased safety concerns and legislatively mandated changes in industrial and automotive policies.

Lunde and Bjorseth (1977) compared samples of air with trajectories from northern Norway against those with trajectories from the more industrialized southern Norway; each of 20 polycyclic aromatic hydrocarbons ranged from less than 10 picograms to more than 500 picograms per cubic meter and were significantly higher in the air originating in the more industrialized sections of southern Norway. This study is particularly significant because it shows that at least these 20 chemicals are stable enough to be transported from their industrial source to human population densities where the chemicals can be inhaled by significant numbers of persons.

Polycyclic aromatic hydrocarbon contamination is 5–20 times higher in polluted river and lake waters than that in ground water (Borneff, 1977). Pollution of drinking water will, of course, parallel that found in the body of water from which the drinking water originates (Loper, 1980). Polycyclic hydrocarbon concentrations up to 12 micrograms per liter have been recorded in severely polluted rivers. In polluted rivers used as drinking water sources, benzo[a]pyrene or total polycyclic aromatic hydrocarbon concentrations range from 0.1 to 140 nanograms per liter. Treatment by sedimentation, flocculation and filtration can decrease these levels by as much as 500 to 1,000 fold.

The capacity for benzo[a]pyrene metabolism (*viz.* aryl hydrocarbon hydroxylase activity) in fish and shellfish can be used as a measure of the induction of this enzyme by environmental polycyclic hydrocarbon pollutants and has been suggested (Mix and Schaffer, 1979; Kurelec *et al.*, 1979) as a means of monitoring the degree of polycyclic aromatic hydrocarbon pollution of water.

Vegetables, fruits, and meats produced within a polluted environment can be expected to contain higher concentrations of polycyclic aromatic hydrocarbons than such foods produced within a nonpolluted environment (Combes and Haveland-Smith, 1982). The style of cooking, choice of foods, and seasonal, regional, and personal variations in diet are all factors making measurements of polycyclic aromatic hydrocarbon ingestion extremely difficult to quantify. Various polycyclic hydrocarbon concentrations in raw foods range up to 30 ng per gram. Many smoked and charcoal-cooked meats exhibit polycyclic aromatic hydrocarbon concentrations of 10 to greater than 100 ng per gram (Santodonato *et al.*, 1981). Factors enhancing the amount of polycyclic hydrocarbon residue include the length of cooking time, contact with the flames, and high temperatures. Cigarette, cigar, and pipe smoking (both active and passive smoking) also provide a source of polycyclic organic matter. One pound of well-done charcoal-broiled hamburger contains as much benzo[a]pyrene as 600 smoked cigarettes.

In addition to many organic chemicals, carcinogenic metals such as cadmium (Degraeve, 1981) are present in significant amounts in the environment, particularly near industries using large amounts of such metals.

5.3.3 *Route of Exposure*

Total body accumulation of chemical carcinogens occurs by intake from inhalation, ingestion, and percutaneous absorption. Because of large differences in the diet and environment among individuals, the human daily consumption of carcinogens is highly variable. Estimates of the daily intake range between nondetectable levels and 30 nanograms per kilogram body weight per day (ICRP, 1974). Such estimates are made by measuring the concentrations of several known carcinogens in all kinds of foods (and air and water) and then determining the amounts of these foods ingested per day within the human population. Clearly these values are approximations that can be in error by several orders of magnitude.

Total body accumulation reflects both total intake and the rate of elimination. Factors important in the rate of elimination include pharmacokinetics, lipid solubility, metabolism of the parent compound, profile of metabolites formed, rate of formation of reactive intermediates, degree of enzyme induction, amount of relevant covalent binding with subcellular macromolecules, and the rate of removal from the cell of subcellular macromolecules having covalently bound intermediates. Genetic differences in metabolism constitute a major cause of large variations in elimination rates within the human population. Differences in drug metabolism within the human population easily can vary by 200-fold (Goldstein et al., 1974), and among species they can vary by more than 1,000-fold. Extrapolation of a particular drug effect or its metabolism from one laboratory animal to another, or to the human population, can therefore by very problematic.

5.3.4 Concentrations in the Body

The means for measuring exposure to carcinogenic chemicals are more complicated than those for determining exposure to ionizing radiation for external or even internal exposures. For external ionizing radiation, especially total-body exposure, the mean dosage to a particular tissue can be readily calculated. For chemicals, numerous parameters interact: the type of chemical, whether the metabolite(s) or the parent compound is the ultimate carcinogen, the dose, route of administration, duration of exposure, genetic differences in metabolism, route and rate of excretion, and the number of possible intermediate metabolic pathways (both "activation" and "detoxification"). Ultimately, of course, the probability that a given carcinogen will exert a carcinogenic effect depends on the concentration of the active form of the chemical that exists at the relevant target site, or sites, and the length of time it remains there.

Wearing a standard radiation badge and monitoring one's urine are two means of making estimates of exposures to external and internal sources of ionizing radiation. Fundamental detection methods for exposure to chemical carcinogens, however, include tissue or excreta analysis by solvent partitioning, chromatography, flame ionization and electron caputre, UV absorption, luminescence and sensitized fluorescence, nuclear magnetic resonance, and gas-chromatography/mass spectrometry. The limit of sensitivity for any of these later techniques is in the range of 1 nanogram to 10 picograms.

More sensitive biological assays include testing for the mutagenicity of carcinogenic chemicals in body fluids: urine (Bloom, 1981), feces (Bruce et al., 1977) or blood (Legator et al., 1975; Dobias, 1980). The use of adducts in hemoglobin from red blood cells as a biological dosimeter has been explored; measurement of alkylated amino acids in hemoglobin from exposed individuals is not a genetic alteration, but protein binding strongly implies concurrent alkylation of DNA (Ehrenberg, 1979). Detection of 6-thioguanine resistance in peripheral blood lymphocytes (Strauss, and Albertini, 1979; Albertini and Allen, 1981) and of variant hemoglobins (Mendelsohn et al., 1980) have been proposed as sensitive techniques for detecting somatic mutations in individuals exposed to mutagenic carcinogens. However, studies utilizing these endpoints are exceedingly few (Bloom, 1981). Determination of the rate of unscheduled DNA synthesis (UDS) in human lymphocytes (Regan and Setlow, 1973; Cleaver, 1977) also has been suggested as an indicator of exposure of chromosomes to DNA-damaging agents. Chromosomal aberrations and sister chromatid exchanges (SCE) are used routinely as *in vivo* indicators of exposure to genotoxic chemicals (Stetka and Wolff, 1976a; Kucerova et al., 1979; Nakanishi and Schneider, 1979; Bloom, 1981; NAS/NRC, 1983a; Tice, 1984; Tice and Hollaender, 1984b), in *in vitro* assays to detect mutagens (Stetka and Wolff, 1976b; Carrano et al., 1978; Gomez-Arroyo et al., 1981). However, for several of these endpoints, e.g., SCE, there is no clear-cut indication of how closely the endpoint may be associated with mutagenesis or carcinogenesis (Jostes, 1981; Latt et al., 1981).

The assessment of DNA adducts may provide a sensitive indicator of previous exposure. The enzyme-linked immunosorbent assay (ELISA) has a lower limit of detection of about 0.08 femtomol per microgram of DNA (Perera et al., 1982). This assay requires (1) the development of an antibody specific for a certain chemical metabolite bound covalently to DNA and (2) the isolation of DNA from some tissue sample, e.g., skin biopsy, or lymphocytes of an exposed individual. It is anticipated that further refinement of such immunologic techniques may lower the threshold of sensitivity by one or two orders of magnitude. One such refined test is the ultrasensitive enzymatic radioimmunoassay (USERIA), purported to be about five times more sensitive than ELISA (Hsu et al., 1981; Shamsuddin et al., 1985; Harris et al., 1985). Quantification by the development of monoclonal antibodies to aflatoxin B_1 metabolites bound to DNA (Groopman et al., 1982; Sizaret et al., 1982) has now been reported.

Due to genetic differences in metabolism and enzyme induction, the persistence of a chemical or its metabolite can vary widely among different tissues within an organism, among individuals and among species. Variations in radiation effects among different exposed tissues are not as striking as the variations in chemical effects among different tissues. The inherent chemical structure and stability of each carcinogenic molecule is also important. The particular profile of metabolites (both reactive and inactive) can vary widely due to genetic differences in metabolism and enzyme induction (Thorgeirsson and Nebert, 1977). An additional parameter to consider is the persistence of reactive intermediates that bind covalently to nucleic acids or proteins. It is well known that DNA damage is detected and repaired in eukaryotes by the appropriate DNA repair systems (Lindahl, 1982; Loeb and Kunkel, 1982); such damage includes the covalent binding of active chemical intermediates to DNA. RNA and proteins that are defective (such as those having covalently bound metabolites) do not survive as long as their nondefective counterparts. Moreover, damage induced by different metabolites or different kinds of damage induced by the same substance may be detected differentially. For example, the 4,5-oxide of benzo[a]pyrene covalently bound to DNA is removed more rapidly than the covalently bound 7,8-*trans*-diol-9,10-epoxide of benzo[a]pyrene (Pelkonen et al., 1979; Feldman et al., 1980; reviewed in Pelkonen and Nebert 1982).

One of the most fundamental problems is to determine which test reflects the best assessment of the concentrations of carcinogenic chemical species in the body. Some advocate merely the total dose of chemical detected per gram of tissue, or per milliliter of urine or feces. Others believe that the binding of a carcinogen (usually a metabolic intermediate) covalently to DNA (or RNA or protein) or the degree of damage to cells resulting from the exposure provides a more accurate assessment. This latter measurement certainly reflects consideration of such parameters as genetic differences in carcinogen metabolism, metabolite profile, rates of detoxication, and even proficiency of DNA repair. There are known carcinogens, however, which require metabolic conversion to reactive derivatives, or intermediates, and these intermediates cause free radical formation, which in turn can damage DNA, e.g., hydroxymethyluracil formation. In this instance, "covalent binding of a carcinogenic metabolite to DNA" may not occur at all in spite of sufficient damage of DNA to initiate carcinogenesis. Much research, therefore, is still necessary to assess "carcinogenic potential" for any given chemical in any given tissue.

5.4 Pharmacokinetics

There are almost no data available concerning the pharmacokinetics (i.e., the uptake, distribution, metabolisms, and excretion) of chemical carcinogens in humans. Nevertheless, it is possible to make limited assumptions about the pharmacokinetics of carcinogens, based on the results of animal studies conducted with various chemicals, notably polycyclic hydrocarbons such as benzo[a]pyrene.

5.4.1 Uptake

The fact that chemical carcinogens can cause toxicity or tumors in distant tissues of laboratory animals following oral, inhalation, or dermal administration supports the thesis that these chemicals can cross epithelial membranes. Polycyclic hydrocarbons (Nebert and Bausserman, 1970) or 2,3,7,8-tetrachlorodibenzo-*p*-dioxin (Okey et al., 1980) readily enter the cytoplasm and nuclei of cultured cells held at 4 °C. Tissue accumulation of benzo[a]pyrene throughout the intact animal is exponential as oral benzo[a]pyrene concentrations are increased (Rees et al., 1971). All these data are consistent with a mechanism involving *passive diffusion* of such chemical carcinogens into cells. The fact that chemicals given to pregnant laboratory animals can cause tumorigenesis in the offspring months or years after birth (Tomatis, 1965) indicates that this passive diffusion process includes transplacental passage of chemical carcinogens.

5.4.2 Distribution

Regardless of the route of administration, chemical carcinogens, once absorbed, become distributed throughout the body. Obviously the absorbed chemicals will follow normal blood or lymphatic movement. Absorption from the peritoneal cavity via the mesenteric blood vessels results in movement of the chemicals through the liver before reaching the heart. Absorption from the lungs results in transfer of the chemicals via the left side of the heart to all peripheral tissues. Generally the more fat-soluble the molecule, the greater is the concentration found in adipose tissue and secreta such as breast milk, sebum from sebaceous glands and semen. Transplacental passage of chemical carcinogens occurs as readily as with drugs of similar hydrophobicity. Particulate materials such as asbestos or polycyclic organic matter, especially in the lungs, may be phagocytosed and long-term holding of particulate matter containing absorbed polycyclic hydrocarbons has been postulated as a mechanism for exposing Clara and alveolar Type II cells to quantities sufficient to initiate lung cancer (Kauffman et al., 1979; Sato and Kauffman, 1980).

5.4.3 Metabolism

Before 1960 it was generally believed that most drugs and other environmental chemicals were pharmacologically active, toxic, carcinogenic, and/or mutagenic in their parent (nonmetabolized) form. The function of drug-metabolizing enzymes was, therefore, regarded as *detoxication*, i.e., to "inactivate" the active parent drug. Now it is evident that, although some chemicals are indeed active in their nonmetabolized form, many foreign compounds are inactive until metabolized; the process through which active derivatives are formed is called *toxification*, or metabolic activation. Detoxication and metabolic activation enzymes coexist in the same cell. Further, any given enzyme may be involved in detoxifying one chemical while toxifying a second chemical. Examples are given later in this section. Each enzyme is controlled by one or more genes, and more than 300 agents are known to induce their own metabolism or the metabolism of chemically related compounds (Nebert *et al.*, 1981). Often the level of a particular enzyme differs among various tissues or organs, strains, and species and whether the cells are *in vitro* or *in vivo*. Factors such as age, hormonal and nutritional variations, diurnal and seasonal rhythms, pH at the enzyme active-site, and saturating versus nonsaturating substrate concentrations also contribute to the function of each of these enzymes. A delicate and very complicated balance exists, therefore, between detoxication and toxification.

In general, changes in the metabolism of drugs or other chemicals as a function of time are negligible during embryogenesis, increase during the fetal and neonatal periods to a peak at the time of weaning, and then decrease gradually during the rest of the life of the laboratory animal. Human studies, e.g., of antipyrine or caffeine clearance in individuals between the age of 16 and 100 years, are consistent with the general pattern of drug metabolism described above for laboratory animals. Except during the gestational and neonatal periods of laboratory animals, therefore, drug metabolism (and inducibility) varies at most only a few-fold between the weaning period and old age. Thus, *genetic* differences in the metabolism of drugs or other chemicals—which vary by factors of 20-200 fold between individuals (Gonzalez *et al.*, 1986) and even more than that between species—are usually more important in determining the rate of metabolism of such chemicals than age.

Increases in carcinogenesis are observed under certain experimental conditions. The embryo is highly sensitive to some types of carcinogens, and tumorigenesis in the liver occurs after partial hepatectomy much more readily than in the absences of liver damage and regrowth. Furthermore, ethylnitrosourea-induced tumors occur in the brain, where repair of ethylnitroso-urea-DNA adducts is poor, and not in liver, where repair of such adducts is efficient. Thus with many cancers, increases in carcinogenesis reflect not only differences in drug metabolism but also other differences in the host, *viz.*, differences in cell turnover, DNA repair, immune response, etc. It must be stressed, therefore, that metabolism per se does not fully explain differences in the way cancer is initiated, promoted, and finally expressed.

Many environmental pollutants are so fat-soluble that they would remain in the body indefinitely were it not for the metabolism resulting in more water-soluble derivatives. These enzyme systems, located to some degree in virtually all tissues of the body, are usually divided into two groups: Phase I and Phase II. During Phase I metabolism, one or more water-soluble groups (such as hydroxyl) are introduced into the fat-soluble parent molecule, thus allowing a "handle," position, or functional group for the Phase II conjugating enzymes to attack. Many Phase I products, but especially the conjugated Phase II products, are sufficiently water-soluble so that these chemicals are excreted readily from the body. Because there are several examples of Phase II reactions preceding Phase I reactions, Testa and Jenner (1978) recommended use of the term "functionalization reactions" instead of "Phase I reactions." In like manner, "conjugation reactions" might be a better term for "Phase II reactions."

Listed in Table 5.3 are a number of metabolic reactions that may be important in detoxication or toxification of mutagens and promutagens (see also Jakoby et al., 1982). Basically, any time a chemical bond is cleaved and/or electrons are passed one at a time, the possibility exists for unwanted reactions of such intermediates with nucleic acid or protein. The reactions can be complicated, depending on the degree of stability of short-lived chemical intermediates, the redox state, movement of unpaired electrons from one molecule to another (free radicals), and lipid peroxidation. Generally, for mutagenesis and carcinogenesis, reactions involving DNA are believed to be most important, but absolute experimental proof of this hypothesis is still lacking. It is certainly possible also that mutagens can interact with nucleic acid polymerases, endonucleases, RNA splicing and processing enzymes, or other important proteins, thereby leading to heritable changes in gene expression.

The majority of all Phase I oxidations is performed by cytochrome P-450. "P-450" denotes a reddish pigment with the unusual property, when reduced and combined with carbon monoxide, of having its major optical absorption peak (Soret maximum) at about 450 nm. Although the name P-450 was intended to be temporary until more knowledge about this substance was known, the terminology has persisted for 25 years because of the complexity of this enzyme system and the resulting lack of agreement on any better nomenclature.

Cytochrome P-450 represents a huge family, containing 30 to 200 genes (Gonzalez et al., 1986) encoding hemoprotein enzymes that possess catalytic activity toward thousands of substrates. These enzymes

TABLE 5.3—*Potentially important metabolic reactions that may play a role in environmental mutagenesis and carcinogenesis*

A. Oxidations (Phase I metabolism)
   1. Aromatic or aliphatic C-oxygenations (epoxidations, hydroxylations)
   2. N-, O, or S-dealkylations
   3. N-oxidations or N-hydroxylations
   4. S-oxidations
   5. Deaminations
   6. Dehalogenations
   7. Metallo-alkane dealkylations
   8. Desulfurations
   9. Alcohol or aldehyde dehydrogenations
   10. Xanthine (and other purines) oxidations
   11. Tyrosine hydroxylation
   12. Monoamine (including catecholamine) oxidations

B. Reductions (Phase I metabolism)
   1. Azo reductions
   2. Nitro reductions
   3. Arene oxide reductions
   4. N-hydroxyl reductions
   5. Quinone reductions
   6. Carbonyl sulfide reduction by carbonic anhydrase
   7. Reductions of transition metal ion salts

C. Hydrolyses (Phase I metabolism)
   1. Hydrolyses of esters
   2. Hydrolyses of amides
   3. Hydrolyses of peptides

D. Conjugations (Phase II metabolism)
   1. Glucuronidations
   2. Sulfate conjugations
   3. Glutathione conjugations
   4. Acetylations
   5. Glycine conjugations
   6. Serine conjugations
   7. N-, O-, or S-methylations
   8. Ribonucleoside or ribonucleotide formations
   9. Glycoside conjugations
   10. "Water conjugations"

E. Beyond Phase II metabolism
   1. C-oxygenations
   2. Glucuronidations
   3. Glycosidations
   4. Deacetylations

F. Chemical reactions (oxidation/reduction)

are known to metabolize: many drugs and laboratory reagents; small chemicals such as benzene, thiocyanate, or ethanol; polycyclic aromatic hydrocarbons such as benzo[a]pyrene, halogenated hydrocarbons such as polychlorinated and polybrominated biphenyls, herbicides, insecticides, and ingredients in soaps and deodorants; certain fungal toxins and antibiotics; many of the chemotherapeutic agents used to treat human cancer; strong mutagens such as nitrosamines; aminoazo dyes and diazo compounds; many chemicals found in cosmetics and perfumes; numerous aromatic amines, such as those found in hair dyes, nitro aromatics, and heterocyclics; N-acetylarylamines and nitrofurans; wood terpenes; epoxides, carbamates; alkylhalides; safrole derivatives; antioxidants, other food additives, and many ingredients of foodstuffs, fermented alcoholic beverages, and spices; both endogenous and synthetic steriods; prostaglandins; and other endogenous compounds such as biogenic amines, indoles, thyroxine, and fatty acids. The growing consensus among scientists in the field is that less than 200 P-450 genes exist and that overlapping substrate specificity accounts for all diversity seen when thousands of different chemicals are metabolized. There is no experimental evidence to date that more than one P-450 protein is derived from a single gene (Nebert and Gonzalez, 1985).

The various forms of P-450 represent a large subset of all monooxygenases, which are enzymes that insert one atom of atmospheric oxygen into their substrates. P-450-mediated monooxygenases are ubiquitous in virtually all living things—certain kinds of bacteria, and presumably all plants and animals. For example, some types of bacteria are being developed to destroy oil spills in the ocean; such a catalytic activity includes Phase I metabolism and the P-450-mediated monooxygenase system.

P-450-mediated monooxygenase activities (Table 5.3) include: aromatic and aliphatic hydroxylations of carbon atoms; N-, O-, and S-dealkylations; N-oxidations and N-hydroxylations; S-oxidations; deaminations; dehalogenations; metallo-alkane dealkylations; desulfurations; certain purine and monoamine oxidations; azo and nitro reductions; and certain arene oxide and N-hydroxyl reductions. Removal of methyl or ethyl groups from substrates (dealkylations) can result in the covalent binding of reactive intermediates (alkylation) to nucleic acid or protein, a potentially important mechanism for mutagenesis, tumorigenesis, or drug toxicity. Chromate (Cr VI) is one of a number of inorganic chemical carcinogens. P-450 has been reported to reduce *in vitro* certain metals such as Cr(VI) to Cr(III); this, therefore, may be an important detoxication pathway for chromate.

Human liver alcohol dehydrogenase (Table 5.3) catalyzes the oxidation of digitalis derivatives. This is a potentially important *detoxication* pathway. Liver alcohol dehydrogenase may also *toxify* chemicals, metabolizing several xylyl alcohols to aldehydes that are toxic to lung cells and allyl alcohol to the extremely neurotoxic acrolein. Alcohol dehydrogenase is therefore an excellent example of the dual nature of an enzyme designed to metabolize endogenous substrates: the enzyme is capable not only of detoxifying one foreign chemical but is also capable of toxifying another.

Quinone reduction by DT diaphorase has been postulated to be an important step leading to glucuronide conjugation (Table 5.3). Quinone-derived free radicals also might be generated by this catalytic activity or by some similar activity other than DT diaphorase. Carbonyl sulfide has recently been shown to be metabolized to hydrogen sulfide by carbonic anhydrase; hydrogen sulfide is responsible for carbonyl sulfide toxicity. Carbonic anhydrase therefore represents a metabolic pathway leading to toxification.

Drugs and other foreign chemicals are conjugated most commonly with glucuronide, sulfate, or glutathione (Table 5.3). Studies have only begun on characterization of the sulfotransferases. Glutathione transferases act on a large number of chemicals—including arene oxides, epoxides, chlorodinitrobenzene, bromosulfophthalein (for testing liver function), and bilirubin; at least eight glutathione transferases have been characterized so far. Epoxide hydrolase adds water to arene oxides or epoxides to form dihydrodiols. This "water conjugation" occurs, for example, during the metabolism of carcinogenic polycyclic hydrocarbons.

Although dihydrodiols are readily excreted, in general, it is now clear that diol-epoxides are formed and that these highly reactive intermediates may be important in mutagenicity, tumorigenesis, toxicity, and birth defects. Diols can therefore undergo further C-oxygenations (Table 5.3).

Once a conjugate is formed, the general belief has been that the drug is excreted irreversibly. However the glucuronide of 3-hydroxybenzo[a]pyrene treated with β-glucuronidase activity can lead to high levels of covalent binding of reactive intermediates in rat kidney and bladder. By a similar mechanism, glycosides react with various glycosidases, resulting in reactive intermediates capable of binding covalently with nucleic acid and protein. Following conjugation with acetic acid, many drugs can be deacetylated (Table 5.3) to form reactive intermediates.

Some chemicals by their inherent molecular properties possess a high redox potential. o-Aminophenol, for example, is capable of oxidizing ferrohemoglobin to ferrihemoglobin. Nitrates also cause methemoglobinemia. Such one-electron chemical reactions may play a role in mutagenesis.

Perhaps only a few enzymes (or none) exist in the body solely to take care of foreign chemicals. Many enzymes designed for normal-body substrates, however, apparently are capable of interacting with most mutagens and promutagens. the result is a complicated and delicate balance of detoxication and toxification, which may be highly dose-dependent (see for example, Dietz et al., 1983).

When patterns of chemical carcinogen metabolism in the bronchus, esophagus, colon and duodenum of the same individual have been examined (Autrup et al., 1982), 30-fold intra-individual variations have been found. No correlation has been noted between cigarette smoking history and the quantity of benzo[a]pyrene-modified DNA found by either the ELISA (Perera, 1981) or the USERIA (Hsu et al., 1981) techniques in these tissues. Such data point out the large degree of tissue variability and especially genetic variability of drug metabolism within the human population. It now seems clear that at least part of the individual increased risk of bronchogenic carcinoma caused by cigarette smoking is correlated with genetically predetermined highly inducible benzo[a]pyrene metabolism measured in cultured lymphoblasts of these patients (Kouri et al., 1982). The human P-450 gene responsible for this metabolism has been cloned and characterized (Jaiswal et al., 1985a).

5.4.4 *Excretion*

Just as most chemical carcinogens are passively absorbed, so are they passively excreted, through the hepatobiliary system and feces, and also via urine, sweat, semen, menses, and milk. The route of administration and its influence on the rate of elimination are likely to be important determinants (Aito, 1974). If the chemical is broken down to small volatile compounds, its excretion from the lung may occur via exhalation; otherwise, a carcinogen in the airway may be cleared via the mucociliary "elevator" to the pharynx and then swallowed, or it may enter the blood stream and be transported to distal tissues. Carcinogen excretion in laboratory animals follows patterns similar to those seen with drugs: an alpha phase during which the chemical peaks; a beta phase of rapid elimination; and a gamma phase of slow elimination during which the "deep compartment," such as adipose tissue, comes into play (Routledge and Shand, 1979).

5.4.5 *Route of Administration*

The pharmacokinetics of benzo[a]pyrene have been studied among mice differing at a single genetic locus (Nebert, 1981). The studies indicate the important interplay among the route of administration, "first-pass" elimination kinetics, and genetic differences in drug metabolism due to enzyme induction controlled by a receptor.

The Ah locus controls the induction of a particular subset of drug metabolizing enzymes by numerous polycyclic aromatic compounds. A cystosolic receptor highly specific for these inducers has been shown to be essential for the induction process. Mice having high levels of the high-affinity receptor (Ah^b/Ah^d) therefore have drug-metabolizing enzymes easily inducible by these polycyclic aromatic compounds. Mice having the poor-affinity receptor (Ah^d/Ah^d), on the other hand, have drug-metabolizing enzymes that are more difficult to induce with these same polycyclic compounds.

From numerous studies of various types of tumorigenesis and toxicity, a pattern has emerged with the use of these mice (Figure 5.1). If the chemical administered to intact mice is an inducer of forms of P-450 controlled by the Ah receptor, not only are the size and timing of the dose important, but the route of administration determines the site at which the tumor or toxicity occurs. In mice having the high-affinity

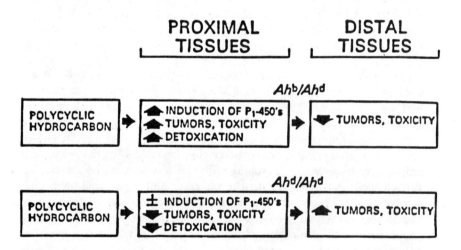

Fig. 5.1 Diagram of the influence of genetic differences in enzyme induction on metabolism, pharmacokinetics, and ultimate site of tumorigenesis or toxicity of a chemical. Ah^b/Ah^d, mouse having sufficient amounts of high-affinity Ah receptor, thereby allowing with ease the induction of cytochromes P_1-450 by polycyclic hydrocarbons. Ah^d/Ah^d, mouse having Ah receptor defect, thereby allowing only with difficulty the induction of cytocromes P_1-450 by this class of chemical carcinogen. (Arrows pointing upward indicate an increase and those pointing down indicate a decrease.)

receptor, polycyclic aromatic compounds applied topically, subcutaneously, or intratracheally cause more tumors or toxicity in tissues at the site of administration: epidermal carcinoma or ulceration, subcutaneous fibrosarcomas, and various types of pulmonary tumors respectively. When administered intraperitoneally to high-affinity receptor mice; these compounds cause more hepatic necrosis and ovarian toxicity than observed in poor affinity mice; again, liver and ovary are viewed as intraperitoneal organs close to the site of the administered drug. On the other hand, in mice having the poor-affinity receptor, polycyclic aromatic compounds given orally or subcutaneously cause aplastic anemia, leukemia, and lymphatic tumors more frequently. These malignancies and toxicity are manifest in tissues distant from the site of the administered drug. In the example of oral benzo[a]pyrene, pharmacokinetic studies have shown a 10 to 20-fold higher uptake of benzo[a]pyrene in the marrow and spleen of poor-affinity receptor mice (Nebert et al., 1980), thus confirming the phenomenon called "first-pass elimination" kinetics (Routledge and Shand, 1979). "First-pass elimination" kinetics can be defined as follows: if a chemical administered by any route is metabolized to ineffective levels before it reaches its target tissue, it will cause no response in that tissue. Hence, the forms of P-450 induced in the intestine and liver efficiently metabolize the benzo[a]pyrene given orally to the mouse having the high-affinity receptor and the amount of parent drug and/or its metabolites reaching the bone marrow is therefore less than in the poor-affinity receptor mouse.

If the compound induces its own metabolism at the site of administration to a greater degree in high-affinity receptor mice than in poor-affinity receptor mice, the former may have higher local levels of reactive intermediates which can cause tumor formation or toxicity at that site (see Figure 5.1). On the other hand, if the compound induces detoxication pathways also at the site of administration to a greater degree in high-affinity receptor mice, malignancy or toxicity may occur in tissues *distal* from the site of administration. Because the concentration of nonmetabolized parent drug that reaches distal tissues (such as bone marrow and lymph nodes) may, therefore, be higher in poor-affinity receptor mice, distal toxicity or malignancy are also likely to be greater in such animals. Although no similar pharmacokinetic data in man are known, it is clear that humans have the Ah receptor (Nebert, 1981; Jaiswal et al., 1985b). The human P_1-450 gene (that P-450 which is correlated with inducible polycyclic hydrocarbon metabolism) has been isolated and sequenced (Jaiswal et al., 1985a) and mapped to chromosome 15 (Hildebrand et al., 1985). In addition to the induction of enzyme activity, carcinogens may also saturate or inhibit enzyme activity in some instances, the nature and extent of reaction varying with the dose and conditions of exposure.

It should be emphasized, however, that this relationship for genetically determined enzyme toxicity has been worked out in detail only with polycyclic hydrocarbon carcinogens. It remains to be determined how many other classes of carcinogens may also show such differences.

If the mutagenicity of a chemical is determined with the use of the *Salmonella*/liver microsome test, the results depend on toxification of the chemical by the postmitochondrial supernatant fraction from livers of rats previously treated with Aroclor 1254. Likewise, any other mutagenesis index for a given chemical requiring metabolism depends solely on the drug-metabolizing enzyme systems that are used for toxification and the *balance* of toxification-to-detoxication in the particular form of P-450 that enhances the toxification of that chemical, for example, the tumorigenicity of the chemical need not be correlated at all with mutagenicity determined by the *Salmonella*/liver microsome test if the same form of P-450 in mouse skin is not present in the rat liver fraction. In the intact animal, therefore, drug-drug interactions, chemicals which promote cancer, and agents that enhance carcinogenesis may act in concert in a manner that is not possible to appreciate in any cell culture or *in vitro* test system. For this reason, a mutagenicity assay may give false negative, as well as false positive results.

5.5 Summary

In contrast to penetrating radiation, which passes through tissue in a well-known manner, or even with the more complex situations arising from nonuniform distributions of internally deposited radionuclides, dosimetry of chemicals is far more difficult. Factors requiring consideration in "chemical dosimetry" include: the type of chemical under consideration, dose, duration of exposure, route of administration, all possible metabolic pathways, capacity of the chemical to affect its own metabolism, pharmacokinetics, rate of excretion, and dose to the biological target at sites of tumor formation (which may vary depending on the route of administration).

The complexity of the enzyme systems responsible for metabolizing chemicals involves an admixture of cascading pathways, routes of detoxification and metabolic activation, feedback inhibition, and possibilities for enzyme induction. Many of the individual pathways are controlled by separate genes, and each human (except an identical twin) is genetically unique. It is clear, therefore, that extrapolation of laboratory animal data to the human is extremely difficult. With regard to rates of metabolism of chemical carcinogens, more than four orders of magnitude of variation have been observed among non-human species, and variation of more than two orders of magnitude are to be expected within the human population. Although policies for limiting exposure to carcinogens are formulated on the basis of the average human, genetically determined individual differences in metabolism are large. These account for much of the variation in response to chemicals that is seen within the human population.

6. Carcinogenic Effects of Radiation and Chemicals in Man

6.1 Radiation

Human data on the carcinogenic effects of ionizing radiation have been obtained from studies of patients exposed to diagnostic or therapeutic radiation, survivors of A-bombs dropped at Hiroshima and Nagasaki, and workers exposed to a wide variety of radiation hazards (UNSCEAR, 1977; NAS/NRC, 1980). Although subject to considerable uncertainty, risk estimates based on these data are useful in the determination of radiation protection policy, in making decisions relative to compensation for radiation-induced injury, and in the effort to understand the biology of radiation carcinogenesis and, perhaps, even of carcino-genesis generally.

The accumulation of human data on a wider scale, and in quantitative form, has enabled radiation epidemiology to advance from an essentially descriptive stage to an analytic one in which numerical, dose-specific, risk estimates have begun to be treated statistically in order to identify determinants of risk. As presently conceived, these determinants include characteristics of the radiation exposure, underlying dose-response relationships, host factors, other environmental factors, differential tissue sensitivity, time after exposure, and natural levels of incidence.

6.1.1 *Characterization of Exposure*

The modalities of exposure for which there are at least some human data include the part of the body exposed to radiation, the dose to an organ or tissue, the distribution of the dose over time (dose-rate and fractionation), and the radiation quality defined in terms of LET.

6.1.1.1 *Part of Body Exposed*

Diagnostic and therapeutic radiation involves almost exclusively partial-body exposure, in contradistinction to that received by A-bomb survivors and by workers with external exposure. Except for bone sarcoma, tissue-specific risk estimates based on partial-body exposure are reasonably consistent with those based on whole-body exposure, but this may be only because such estimates are not yet made with sufficient precision to reveal differences. According to the report of the NAS/NRC Committee on the Biological Effects of Ionizing Radiation (BEIR), (NAS/NRC, 1980), radiation-induced bone sarcomas tend to occur at the skeletal locations of highest risk to naturally occurring tumors; for this reason, localized x-ray therapy which usually does not include such sites may be much less carcinogenic than exposure to bone-seeking radionuclides that are deposited throughout the entire skeleton.

That radiation exposure of one part of the body can affect the carcinogenic risk of a distant part (abscopal effect) has not been shown in man, but the possibility of such an effect is suggested by the observation that sterilization of young women may protect against breast cancer; that is, pelvic irradiation sufficient to greatly diminish estrogen levels might have a similar effect.

6.1.1.2 *Dose to Organ or Tissue*

The dose absorbed by the target organ or tissue is the most important single determinant of radiogenic risk. Estimates of absorbed dose are given as the mean energy absorbed per unit mass of tissue; 1 Gy = 1 J kg^{-1}, and 1 rad = 0.01 J kg^{-1}. The cell transformation that leads eventually to cancer appears to involve only a particular subset of cells. With some exposures, particularly with internal emitters, the dose to that active subset of cells may differ markedly from the mean integrated dose to the entire organ. This seems to be the case, for example, in bone sarcoma induced by radium-224 where the average dose to endosteal tissue is 7.5 times that to the entire skeleton, and the risk coefficient for endosteal tissue is correspondingly different (NAS/NRC, 1980). Finally, the dose estimate is an average value for a particular organ or tissue, independent of the mass or volume of the target tissue. This means that current risk coefficients, e.g., those of UNSCEAR (UNSCEAR, 1977) or BEIR III (NAS/NRC, 1980), incorporate any influence that organ size may itself have on the carcinogenic risk. Under the concept of a mean tissue dose, a partial-body exposure can be expressed in terms of a mean marrow dose for the entire skeleton. Or, if only one breast is irradiated, a mean breast dose can be estimated for both breasts although the other was spared. Size or mass of organ cannot be a major factor in the determination of risk, however, or else the coefficients for bone, liver, and brain would be quite large, and that for the thyroid, quite small.

6.1.1.3 *Dose Rate and Fractionation*

Although the animal data suggest that the interval of time during which a given dose is administered greatly influences the carcinogenic effect, especially with low-LET radiation (NCRP, 1980), the few available observations on man are equivocal. It may also be that there are too many other factors confounding the possible comparisons. There are no stable risk estimates for human cancer arising from continuous low-LET irradiation in the occupational environment or in geographic areas of high natural background. The best human data are derived from epidemiologic studies showing an increased incidence of breast cancer in A-bomb survivors exposed at a very high dose-rate, tuberculous patients treated by pneumothorax and monitored by fluoroscopy every 12 days on the average, and mastitis patients who received one to four x-ray treatments, each separated by one or more days. Parallel analyses of these three series yield absolute risk estimates (excess cancers per million per year per rad) that are remarkably similar by age and time after exposure. There are however, other differences among these three series: background level of natural incidence, radiation quality, and whole-versus partial-body exposure. If relative risk were constant, the very low background rate of breast cancer in Japan would be expected to lead to a low absolute risk estimate in comparison with the other two series. The fact that the absolute risk estimate for Japan and the U.S. are equivalent could, therefore, reflect the influence of dose-rate. Against this interpretation, however, is the fact that risk coefficients are similar for the tuberculosis and mastitis series which have the same level of background incidence but differ markedly in dose fractionation.

For leukemia, the linear risk-coefficient is on the order of two excess cases per million persons per rem per year for A-bomb survivors (NAS/NRC, 1980), 0.5 or 2.0 for ankylosing spondylitis patients, depending on the model (Smith and Doll, 1982) and perhaps one or less for US radiologists (NAS/NRC, 1972). In the period 1950-1974, when the leukemogenic effect of exposure to radiation from the atomic bombs of 1945 was at its peak, individuals exposed to 4+ Gy (400+ rad) had 50 times the risk of leukemia experienced by individuals exposed to less than 0.1 Gy (10 rad) (Beebe *et al.*, 1978a). The A-bomb survivors experienced an almost instantaneous exposure, the ankylosing spondylitis patients a dose in many fractions, none small, and the early radiologists an exposure extending over many years but in an unknown pattern. Although the dosimetry for U.S. radiologists is unsatisfactory, one might think that the data would err chiefly from neglect of some high exposures involving individuals who were careless of their own safety in the reference period 1935-1948, correction for which would only reduce the risk estimate. A comparison of the lower dose-rate data of U.S. radiologists and the higher-dose-rate for the other series would suggest that dose rate might affect the risk of leukemia in man by a factor of two. The data are too imprecise, however, for this to be a firm conclusion.

6.1.1.4 Quality of Radiation (Linear Energy Transfer)

For only a few cancer sites are there comparative risk estimates for radiations of different qualities, and only for lung cancer are there strong indications that the carcinogenic risk depends on radiation quality (NAS/NRC, 1980). The relative biological effectiveness (RBE) and dose-equivalent concepts are predicated on the belief, founded in both experimental observations and radiobiologic theory, that the quality of radiation influences risk. The risk for fast neutrons and for protons of energy less than 10 MeV is greater than that for x rays, gamma rays, and beta particles by a factor of 10 originally, but now 20 (ICRP, 1985; NCRP, 1987b). However, the validity of these recommendations has been questioned extensively. On the basis of microdosimetric principles, experimental work, observed differences in leukemia incidence among the two groups of A-bomb survivors (Hiroshima and Nagasaki), and earlier concepts of the dosimetry in Hiroshima and Nagasaki, Rossi argued that the RBE of neutrons for leukemogenesis, relative to low-LET radiation, may exceed 100 at very low doses (Rossi, 1977). It has been suggested that the leukemia cases in the Nagasaki Life Span Study cohort are too few, and possibly non-representative of all leukemia cases among the much larger number of Nagasaki survivors, to support strong conclusions in this regard (Beebe et al., 1978a). In any case, the issue is now moot because, under the new dosimetry that is being developed, the difference in the neutron dose distribution in the survivors of the two A-bombs may be very small (Loewe and Mendelsohn, 1981; RERF, 1987). If neutrons are arbitrarily assigned an RBE of one, the linear coefficient for leukemia over the period 1950-1978 is 1.72 leukemias per million persons per year per rad (NAS/NRC, 1980), in comparison with 2.4 for ankylosing spondylitis patients with mean bone doses of 2 Gy (200 rad) or less (Smith and Doll, 1982), and about one for U.S. radiologists (NAS/NRC, 1972), if any dose-rate effect may be ignored. Estimates of the leukemogenic effect of the continuous alpha irradiation from thorotrast (thorium oxide) are much less satisfactory but are also of the same order (UNSCEAR, 1977; Mole, 1978); however, Mays et al., (1985) point out that the ICRP quality factor of 20 for alpha particles is too high in relation to the data for leukemia in the European thorotrast series.

As a beta emitter, iodine-131 might be expected to have the same carcinogenic effect on the thyroid as x-ray and gamma radiation. Thus far, however, it has not been possible to show that either diagnostic or therapeutic doses of iodine-131 cause human thyroid cancer (Holm et al., 1980a; 1980b). However, the comparison is confounded by any effect that the dose-rate might have on the risk. In a preliminary analysis of the tumor registry data from Hiroshima and Nagasaki for the period 1950-1974, the then-presumed higher neutron component of dose in Hiroshima was not reflected in a higher risk coefficient for thyroid cancer than was obtained for Nagasaki (Beebe et al., 1978b). However, as pointed out above, the doses received by the A-bomb survivors are being reappraised (RERF, 1987).

The risk of radiogenic breast cancer among A-bomb survivors in Hiroshima is not significantly higher than that in Nagasaki but, again, the level of neutron exposure of the Hiroshima survivors is now very much in doubt (Tokunaga et al., 1984). Perhaps the best evidence that LET affects the carcinogenic risk in man is for lung cancer in uranium and fluorspar miners, who, despite their relatively low dose-rate exposure, have risk coefficients 8 to 15 times higher than the ankylosing spondylitis patients who received x-ray therapy (NAS/NRC, 1980). If the comparison is made with A-bomb survivors on the basis of the most recent data (Kato and Schull, 1982), expected changes in dosimetry, and an adjustment for incomplete reporting of lung cancer on death certificates, the average RBE would be very similar, i.e., in the range of 6 to 12. Overall, then, the most that can be said on the basis of the human data alone is that there are definite suggestions that the RBE's for neutrons and alpha particles may be of the same order as those observed in experimental work, but that the data are much too sparse, complicated by difficulties in dosimetry, and confounded by any dose-rate effect, for firm conclusions to be drawn. Until the human data become more nearly adequate, it would seem reasonable to be guided by the experimental data on animals.

6.1.2 Dose-Response Relationships

Theoretical considerations and experimental results have combined to favor a generic dose-response model of the form:

$$F(D) = (a_o + a_1 D + a_2 D^2)\exp(-b_1 D - b_2 D^2) \tag{6.1}$$

where D is dose, $F(D)$ is incidence at dose D, the parameters a_o, a_1, a_2, b_1, and b_2 are positive, and a_0 is the spontaneous incidence of the effect under study (Brown, 1977; Upton, 1977). Except for Baum's suggestion (Baum, 1973) that the response may be a power function of dose, which has received little support, observations on man have not, thus far, been extensive enough to generate alternative dose-response models or concepts governing the dependence of effect upon dose. Nor have they been extensive enough to establish statistically reliable estimates of the parameters of the generic model, which is very flexible and will take many shapes as the values of the parameters change. Although the use of the non-threshold linear model for low-LET radiations continues in matters of radiation protection, it is gradually being displaced in the scientific literature by the quadratic model ("linear quadratic") except in applications involving certain sites of cancer, especially the breast and the thyroid (NAS/NRC, 1980; Rall et al., 1985). The preference shown for the "linear-quadratic" model in both the BEIR III report and the Radioepidemiological Tables (NAS/NRC, 1980: Rall et al, 1985) represents less a formal statistical determination based on the human data than the influence of underlying radiobiological principles and the weight of the more extensive experimental animal data, including the presence of dose rate factors in much of the data.

The BEIR III risk estimates formulated under several dose-response models demonstrate that the choice of the model can affect the estimated excess more than can the choice of the data to which the model is applied. BEIR III estimates of lifetime excess cancer deaths among a million males exposed to 0.1 Gy (10 rad) of low-LET radiation, derived with the three dose-response functions employed in that report, vary by a factor of 15, as shown in Table 6.1 (NAS/NRC, 1980). In animal experiments with high-LET radiation, the most appropriate dose-response function for carcinogenesis is often found to be linear at least in the low to intermediate dose range (e.g., Ullrich and Storer, 1978), but the data on bone sarcomas among radium dial workers are not well fitted by either a linear or a quadratic form. A good fit for these data is obtained only with a quadratic to which a negative exponential term has been added (Rowland et al., 1978).

Increasing knowledge of intracellular repair mechanisms and of genetic deficiencies in repair, such as are seen in patients with ataxia-telangiectasia (Paterson et al., 1979) argues for the view, advanced by Baum (1973), that the distribution of the human population with respect to susceptibility to the carcinogenic effect of ionizing radiation could be such that a pool of susceptible individuals would drive the dose-response curve up at very low doses, sending it above the corresponding linear fit. Such susceptibility has been demonstrated with respect to the carcinogenic effect of UV radiation in patients with xeroderma pigmentosum (Lutzer, 1977), but thus far only sensitivity to cell-killing has been attributed to ionizing radiation.

TABLE 6.1—*Lifetime excess cancer mortality from a single exposure to 10 rad of low-LET radiation per million males, as estimated with the absolute risk projection model and various dose-response models*[a,b,c]

	Excess deaths		
Dose-response model[d]	Leukemia and bone cancer	Other cancers	All cancers
Linear ($y = a + bD$)	566	919	1,485
Linear quadratic ($y = a + bD + cD^2$)	274	421	695
Pure quadratic ($y = a + cD^2$)	35	52	87

[a] NAS/NRC, 1980.
[b] To a population having the age distribution of the 1969-1971 U.S. life table.
[c] 10 rad = 0.1 Gy
[d] D is equal to the dose in rad.

Although non-threshold models are generally preferred, their validity for man is not demonstrable. The best evidence of effects at very low doses comes from the studies of *in utero* exposure (Stewart, 1960; MacMahon, 1962), which are still controversial (MacMahon, 1981; Monson and MacMahon, 1984), and from the Israeli study of thyroid cancer following depilating doses of x rays to the scalp for tinea capitis (Modan *et al.*, 1977). In the latter study, however, there may be doubt as to the accuracy of the dose reconstruction. Data on the Japanese A-bomb survivors give strong evidence of effects below 0.5 Gy (50 rad), especially for breast cancer (Tokunaga *et al.*, 1979; Tokunaga *et al.*, 1984) and leukemia (Ishimaru *et al.*, 1969).

There is reason to assume that dose-response curves for human cancer would generally turn down in the high-dose region, whether because of cell killing, modification of the cellular environment, or other effects, but in few series is this seen (e.g., Rowland *et al.*, 1978). It has been suggested that the absence of leukemia following high doses of x rays and radium in the treatment of cervical cancer (Hutchinson, 1968; Boice *et al.*, 1984), and the absence of thyroid cancer following therapeutic doses of iodine-131 to patients with hyperthyroidism (Holm *et al.*, 1980b), may reflect the cell-killing effect of high doses. The incidence curve for breast cancer in patients with postpartum mastitis treated by x ray shows evidence of a down turn at doses greater than 4 Gy (400 rad) (Shore *et al.*, 1977; 1986), but this is not seen in the larger A-bomb survivor series or in the Massachusetts fluoroscopy series (Land *et al.*, 1980). In the most recent report on the follow up of the British ankylosing spondylitis patients treated with x rays, a better fit for leukemia was obtained with a function having a linear term and an exponential term for cell sterilization than with a linear term alone. The down turn was in the region of 2+ Gy (200+ rad) and the rate above 5 Gy (500 rad) exceeded that for the 2 - 5 Gy (200-500 rad) region (Smith and Doll, 1982). With the elimination of the experience following any second course of x-ray therapy, this series is now somewhat smaller than before, only 35 cases instead of 52, and the better fit with the exponential term may be fortuitous. Interpretation is complicated by inhomogenicity of dose distribution in the marrow of these patients. The much larger A-bomb survivor series, with 180 leukemia deaths, 21 of which were in subjects exposed to 4+ Gy (400+ rad), shows no evidence of down turn, but, of course, tissue doses are not as high in Japanese survivors of the whole-body exposure to radiation from the A-bombs as they are in the ankylosing spondylitis series.

Although the BEIR III Committee (NAS/NRC, 1980) found that cancer mortality data on A-bomb survivors (1950-1974) were not extensive enough to discriminate among the several dose-response models for low-LET radiation derived from the generic model stated above— i.e., linear, linear quadratic, "pure" quadratic, and linear-quadratic with an exponential cell-killing term—a recent analysis of the incidence data for Nagasaki suggests that the "pure" quadratic model gives the poorest fit and significantly so in some analyses (Wakabayashi et al., 1983). In their analysis of the three major breast cancer series, Land et al., (1980) concluded that dose response was approximately linear although some degree of curvature could not be ruled out. Howe has recently suggested, on the basis of the large Canadian fluoroscopy series, that the dose response for breast cancer is nonlinear, perhaps quadratic, but this conclusion may result from the combination of non-homogeneous material from Nova Scotia and the other provinces (Howe, 1984). The probabilities of causation for breast cancer in the Radioepidemiological Tables were calculated on the assumption of linearity of dose response (Rall et al., 1985).

Although absolute (additive) risk models mainly are used in the study of radiation carcinogenesis, relative risk models command increasing attention as the temporal patterns of radiogenic cancers become better established and lifetime projections must be made as a basis for radiation protection. Typically, the analysis of year-of-birth cohorts exposed to radiation will show that relative risk estimates are more stable over time than absolute risk estimates (Kato and Schull, 1982; Rall et al., 1985), suggesting that relative risk projection models may be more reliable than absolute risk projection models when applied to a particular form of cancer in a particular geographic area. But the same relative risk may not describe a given radiogenic risk equally well in areas of different levels of natural incidence. For example, the absolute risk coefficient for breast cancer is about the same for A-bomb survivors as the Massachusetts fluoroscopy patients, but the relative risk per rad among A-bomb survivors is more than three times that among the fluoroscopy patients because the natural incidence of breast cancer is so much lower in Japan (Land et al., 1980). Table 6.2 provides a comparison of the absolute risk estimates presented in Table 6.1 with the parallel relative risk estimates given in the BEIR III Report (NAS/NRC, 1980).

TABLE 6.2—*Estimated lifetime excess cancer mortality from low-LET radiation per million males, as estimated by different dose-response models and risk projection models[a]*

Dose-response model	Excess deaths, all cancers	
	Absolute risk model	Relative risk model
Single exposure of 10 rad[b,c]		
Linear	1,485	4,792
Linear quadratic	695	2,191
Pure quadratic	87	271

[a] NAS/NRC, 1980.
[b] To a population having the age distribution of the 1969-1971 U.S. life table.
[c] 10 rad = 0.1 Gy

6.1.3 Host Factors

Age and sex markedly influence the level of risk from radiation and, in addition, age at exposure governs the temporal pattern in which risk is expressed. There are suggestions that other host factors may also influence the risk of cancer from exposure to ionizing radiation, e.g., hormonal status, immune competence, and genetic constitution, but none of these is well established as a determinant of risk. Much of the evidence for the determinative influence of age at exposure comes from studies of the A-bomb survivors, but Doll was perhaps the first to point out that the leukemogenic effect of radiation was greater in older than in younger adults (Doll, 1970). Studies of the A-bomb survivors, whose ages ranged over the entire lifespan at the time of exposure, have shown that those under age 15 in 1945 also had a higher radiogenic excess of leukemia than those of intermediate age groups (Ichimaru et al., 1978). The influence of age upon the leukemogenic response to radiation is even more complex, extending to the individual types of leukemia and to the temporal patterns in which the radiogenic excess is expressed (see Section 6.1.6). Age-specific absolute risks of mortality from lung cancer, stomach cancer, and all forms of cancer combined except leukemia are higher among older A-bomb survivors than among younger survivors at exposure, but there is a general tendency of the rate of excess mortality for a fixed age-at-death to be higher among the younger survivors once the period of expression has begun (Kato and Schull, 1982). For thyroid cancer, the risk coefficients for the period 1959-1978 are highest among Nagasaki A-bomb survivors exposed in the first two decades of life and lowest for those exposed at later ages (Wakabayashi *et al.*, 1983). Ron and Modan have recently reported a gradient in thyroid risk with age at exposure in Israeli subjects treated with scalp irradiation in childhood for tinea capitis; those under age 6 have more than twice the risk of those aged 9 to 15 at the time of treatment (Ron and Modan, 1984). This interpretation is complicated, however, by a similar age gradient in the dose to the thyroid (Lee and Youmans, 1970). The risk of breast cancer depends also on age at exposure and the highest coefficients are reported for women exposed in the first and second decades of life (Land *et al.*, 1980; Tokunaga *et al.*, 1984). It is unclear how age influences the risk of radiogenic breast cancer in women exposed after age 40. Only in the A-bomb survivor sample are the older ages well represented, and in that sample women aged 40 or older in 1945 have no excess breast cancer (Tokunaga *et al.*, 1984).

Although the carcinogenic effect of irradiation of the fetus, as occurs in x-ray pelvimetry, remains controversial (MacMahon, 1981; Monson and MacMahon, 1984), the weight of the evidence favors the view that the fetus may be highly susceptible to at least the leukemogenic effect of radiation. Relative risks of 150 per Gy (1.5 per rad) have generally been reported from the larger studies, especially for leukemia in the first 10 years of life (Monson and MacMahon, 1984). If a 50 percent increase in leukemia can actually be caused by x-ray doses of a few rad to the fetus, and is not merely the result of factors associated with the decision to perform pelvimetry, then fetal bone marrow may be five times as susceptible as that of children and adults (NAS/NRC, 1972).

Sex differences in risk are much smaller than those associated with age and are seldom reported. Males have a higher absolute risk of leukemia than do females, but a lower risk of thyroid cancer. These differences correspond to differences in natural incidence. For thyroid cancer, the male:female ratio of absolute risk coefficients is 1:2.6, according to the BEIR III Report (NAS/NRC, 1980), but this is close to the ratio of the natural incidence rates, 1:2 in the Third National

Cancer Survey by NCI for 1969-1971 (Cutler and Young, 1975). In the Israeli tinea capitis study, females have a significantly higher radiogenic risk of thyroid cancer than males only in absolute terms (Ron and Modan, 1984; Shore et al., 1985).

Hormonal factors may be involved in the apparent influence of age on the risk of breast cancer, but this has not been shown. Because multiple risk factors have been identified for breast cancer generally, efforts have been made to test for interaction between radiation and reproductive and genetic variables in modifying the chance of radiogenic breast cancer (Nakamura et al., 1977; Shore et al., 1980), but the published series are small. In the first study, on 164 pairs of A-bomb survivors, no evidence of interaction was seen between radiation and variables in the reproductive and marital history; family histories of breast cancer were too few for analysis (Nakamura et al., 1977). In the second study, on the Rochester mastitis series, with 70 cases of breast cancer among 1,564 women, 1 factor among 11 yielded significant evidence of interaction with radiation: whether irradiation was of a primipara. Primiparae had a higher risk of breast cancer than those irradiated after later births even after adjustment for age (Shore et al., 1980).

In addition to hormonal factors, it would seem reasonable to expect that immune competence and genetic factors such as capacity for DNA repair might influence susceptibility to the carcinogenic effect of ionizing radiation, but in neither case is concrete human evidence available. Among subjects in Rochester, New York, who were given x-ray treatment in infancy for thymic enlargement, the Jewish patients (Group C) have consistently shown a much higher risk coefficient than have those of the rest of the sample (Woodard, 1981). The Israeli tinea capitis study also yields a very high risk of radiogenic thyroid cancer, and it is of some interest that it derives largely from immigrants from Morocco and Tunisia; their risk is about 10 times that of subjects from other countries and from Israel itself. Jews from Morocco are known to have a high prevalence of ataxia telangiectasia (AT), and it has been speculated that heterozygotes for AT may be responsible for the high level of risk in the Israeli tinea capitis study (Ron and Modan, 1984). A recent follow-up of more than 9,000 childhood cancers, in which 222 second primary cancers were ascertained, suggests that children with certain childhood cancers are unusually sensitive to the carcinogenic action of ionizing radiation (Tucker et al., 1984). Finally, in the New York University series, skin cancer following x-ray therapy for tinea capitis is seen in white patients but not in blacks (Shore et al., 1984).

6.1.4 Other Environmental Factors

There may well be environmental risk factors that interact with ionizing radiation to cause more excess cancer than would be expected from the two acting independently, but it is mainly for smoking and radon decay products that the question has received serious attention. What at first was thought to be a synergistic enhancement in level of risk for lung cancer (Lundin et al., 1971) now seems to be a true interaction between these two carcinogens (Whittemore and McMillan, 1983); that is, the risk of radiogenic lung cancer depends on the smoking status of uranium miners. On the other hand, studies of Swedish iron miners exposed to radon daughters and followed for long periods (Radford and St. Clair Renard, 1984) provide no evidence of a synergistic effect of smoking in combination with radiation exposure. The initial study of lung cancer in smokers and non-smokers among A-bomb survivors provided no evidence of interaction between smoking and ionizing radiation, but the series was small (Ishimaru et al., 1975). Much larger series have recently been reported, with similar findings (Prentice et al., 1983; Blot et al., 1984). The most suggestive evidence of interaction between ionizing radiation and another environmental carcinogen comes from patients with tinea capitis treated by x rays in whom cancers are observed in areas of the skin of head and face exposed to sunlight but not in areas protected by hair (Shore et al., 1984).

6.1.5 Organ Differences in Response

Organ risks of radiogenic cancer bear almost no relation to their natural incidence of cancer (Table 6.3). This is contrary to the view, often expressed, that the doubling dose of radiation is essentially the same for all cancers. The only exceptions are breast cancer in women and lung cancer in men; in both the natural risk and the radiogenic risk are high. Otherwise a comparison of radiogenic and natural risks is a study in contrasts. Among males, for example, the leading cancers involve the prostate, which appears to be relatively resistant to radiation, the intestines, for which the radiogenic risk is only intermediate, and the urinary organs, for which the radiogenic risk is low. Among women, cancers of the uterus and cervix are second only to breast cancer but the radiogenic risks are too low to have been assessed. Thyroid cancer and leukemia are among the most highly radiogenic forms of cancer but their natural incidence is low (NAS/NRC, 1980). As noted above (Section 6.1.1.2) it seems possible that some of the apparent variation in tissue sensitivity may result from imperfect dosimetric concepts and procedures.

TABLE 6.3—*Average linear risk coefficients for various forms of cancer induced by low-LET radiation in relation to average U.S. incidence rates by sex*[a,b]

Type of cancer	Male Coefficient[c]	Male Incidence[d]	Female Coefficient[c]	Female Incidence[d]
Leukemia[e]	3.1	0.91	2.0	0.57
Thyroid	2.2	0.21	5.8	0.50
Breast	–	–	5.8	7.4
Lung	3.6	7.2	3.9	1.4
Esophagus	0.3	0.57	0.3	0.16
Stomach	1.5	1.5	1.7	0.70
Intestine	1.0	5.2	1.1	4.0
Liver	0.7	0.33	0.7	0.14
Pancreas	0.9	1.2	1.0	0.75
Urinary organs	0.8	3.2	0.9	1.0
Lymphoma	0.3	1.2	0.3	0.79
Other	1.5	13.2	1.6	9.6
All sites	15.9	34.7	25.1	27.0

[a] NAS/NRC, 1980.
[b] Cutler and Young, 1975.
[c] Excess incident cases per million persons per rad per year, age-adjusted.
[d] Cases per 10,000 per year, age-adjusted.
[e] Except chronic lymphocytic leukemia.

6.1.6 *Distribution of Radiogenic Tumors in Relation to Time After Exposure*

There is a characteristic temporal pattern in the occurrence of radiation-induced tumors following irradiation, a pattern that may be segmented into a latency period during which no excess is expressed, a period marked by a rise to a peak or a plateau that may be sustained, and a period of subsidence that may be absent. Only for leukemia and for bone cancer following uptake of radium-224 is this pattern well delineated: a latency period of 2 to 4 years, a peaking at perhaps 6 to 8 years, and subsidence to near zero about 30 years after exposure. For other radiogenic cancers, the minimal latent period is less well defined and is generally taken to be 10 to 15 years.

The degree to which the risks of radiogenic tumors other than leukemia peak rather than plateau is not well established. And whether, as assumed in the BEIR III Report (NAS/NRC, 1980), the expression of excess solid tumors extends throughout life also remains uncertain. The maximum follow-up interval is about 40 years. But observations on the older A-bomb survivors, as well as on other series (Boice and Monson, 1977; Howe, 1984), give no grounds for believing that the period of expression does not persist for the duration of the life span once the latency period has elapsed, at least for certain cancers.

Some uncertainty characterizes the relationship between dose and latent period for radiation-induced cancer in man, for the data are few and somewhat at odds with experimental findings. Animal experiments generally show a shortening of latent period with increasing dose (UNSCEAR, 1977; Mayneord, 1978). Human data are seldom reported but the experience of the A-bomb survivors lends itself to the investigation of this issue, and a recent report (Land and Tokunaga, 1984) provides substantial evidence that the latency of solid tumors may be independent of dose.

One of the most striking aspects of the temporal pattern governing the expression of radiogenic tumors is its dependence upon age at exposure. Excess cases of acute leukemia among A-bomb survivors both began and peaked at progressively later dates for older survivors, and the excess ceased earliest in the youngest subjects, those under 15 years of age of exposure. Although chronic granulocytic leukemia began to appear at about the same time for all age cohorts, those under 15 years of age had the highest peak incidence and their excess ceased sooner than that of other age groups (Ichimaru *et al.*, 1978). For solid tumors the latency period is inversely related to age at exposure and expression characteristically begins only after the exposed group has attained the age at which the particular cancer normally begins to be prevalent. Lung cancer in the A-bomb survivors affords a good example. Those who were at least 50 years of age in 1945 began to manifest their enhanced risk in the mid-1950's. For those who were 35-49 in 1945, the excess seemed to begin about 1960. For those who were 20-34 at exposure the excess first appeared about 1970, and for those who were 10-19, about 1975. Through 1978 there was no evidence of excess lung cancer among subjects less than age 10 years in 1945 (Kato and Schull, 1982).

6.1.7 *Relation of Radiogenic Cancer Incidence to Natural Incidence*

As noted above (Section 6.1.5 and Table 6.3), at the level of organs and tissues, there is no correlation between dose-specific risk estimates and natural incidence. Also, for solid tumors, the minimum latent period is longer for persons exposed at a younger age. The first cancers attributable to an exposure begin only after exposed persons have reached the age at which natural incidence has become appreciable. Whether, once the minimal latent period is passed, the subsequent dose-specific risk for a particular site always follows the level of natural incidence remains unclear, but that it does so is suggested by some of the recent observations on A-bomb survivors (Kato and Schull, 1982; Land and Tokunaga, 1984).

Two models that have been employed for projecting the risks of solid tumors beyond the interval for which observational data exist and for distributing radiogenic cancers over time are: the absolute and the relative risk projection models (Boice et al., 1985a). Both the ICRP and the UNSCEAR present risk estimates in absolute terms. The relative risk model has been used in both the BEIR I and III reports in parallel with the absolute risk model to project risks beyond the interval of observation. The report on the Radioepidemiological Tables (Rall et al., 1985) employs the relative risk model exclusively for all solid tumors except bone cancer; for bone cancer and leukemia, a wave function is used. If complete lifetime observations were available, it might make no difference which model was used: both would, of course, yield the observed excess, although the distribution of the excess over time would probably vary. Since cancer incidence and mortality are strongly dependent upon age at risk, the two models generally give rather different results. The usual pattern is that the relative risk projections exceed absolute risk projections by some margin that depends upon the endpoint in question and the proportion of the lifetime for which observational data exist (Table 6.2).

Only recently have some of the major series on which risk estimates for solid tumors depend been reported for sufficiently long periods of time to permit some examination of the performance of absolute and relative risk models in predicting future risk and in patterning the temporal distribution of radiogenic cancer. Although information is still quite fragmentary, it appears that under certain restrictions the relative risk model fits most observations better and yields more reliable predictions than the absolute risk mode. For example, Kato and Schull (1982), reporting on the mortality experience of A-bomb survivors for the period 1950-1978, show that relative risk estimates for solid tumors within age-at-exposure cohorts are more nearly constant over time than are absolute risk estimates. Unfortunately, they do not provide estimates for individual tumors, but Boice (1979), Land and Tokunaga (1984), and Shore et al., (1986) have indicated that this observation also applies to breast cancer. Similar, but less conclusive, evidence may be seen in the lung cancer experience of the A-bomb survivors. In all of these data, however, it is plain that the relative risk of one age-at-exposure cohort may not be the same as that for another. Also, relative risk coefficients may vary appreciably by tumor type, as between thyroid cancer and breast cancer.

The risk coefficients for breast cancer among the A-bomb survivors, which seem so similar to those for Canadian and U.S. women of the same age at exposure when expressed in absolute terms, are radically different when expressed in relative risk terms. Further, if risk projections are to be made for an undifferentiated set of tumors, as in the BEIR III Report, one must recognize that relative risk coefficients based on observations at the younger ages will have been derived from cancers of quite different types than those to which the coefficients are applied in later life. Examination of the BEIR I Report reveals that about half of the lifetime excess mortality from solid tumors resulting from a lifetime exposure of one rad per year, derives from a risk coefficient for those who were in the first decade of life at exposure and had been followed for only 25 years. Unwilling to project the relative risk coefficient of this young group forward to the end of life, the authors of the BEIR III Report substituted for it the coefficient of those exposed in the second decade of life and followed for 30 years, or until age 40 to 50. They also refused to include in their relative risk estimates certain cancers of later life not known to result from the carcinogenic action of low levels of ionizing radiation: cancer of the prostate, melanoma, and other skin cancers.

In addition to some uncertainty as to the choice of time-response model for projecting risk beyond the period of observation, there are also uncertainties as to the latency period before the expression of radiogenic cancer begins, and as to its duration. For leukemia, these uncertainties are minimal (see Section 6.1.6). But for solid tumors, these parameters are much less well known and may vary appreciably by site. Excess fatal tumors were not seen in the 4-year interval of follow-up, 1975-1978, among A-bomb survivors who were 50 years of age or older in 1945. It is unclear, however whether this is because the carcinogenic effect has run its course, or sampling variation obscures the effect in the very small sample of high-dose survivors remaining alive in 1975. For example, only five deaths were expected among those exposed to 1 Gy (100 rad) in contrast to three observed. An excess was, however, seen in the 1971-1974 period (15 observed vs 9.7 expected). In the preliminary report on the revised and expanded World Health Organization series on cervical cancer patients treated by radiotherapy, there is every indication that the risk of radiogenic solid tumors continues indefinitely, perhaps to the end of life (Boice et al., 1985b).

6.1.8 *Formulation of Dose-Response Models of Radiation Carcinogenesis*

Models are developed as concise formulations of determinate relationships and for purposes of prediction. Their structure derives from mechanistic theory and/or empirical observations manipulated by means of statistical regression models, e.g., polynomials.

Theory, experiment, and empirical observations on man all influence current thinking about models describing the dependence of carcinogenic effects upon radiation dose. It is rather generally agreed among epidemiologists studying the human data that such data are, and will continue to be, an insufficient basis for constructing adequate models or for selecting the most valid model from among alternatives suggested by theory or experiment. There are, however, some propositions that have wide acceptance in the scientific community, viz.: (1) a suitable model is a monotonic increasing function of dose in the dose region of greatest interest, i.e., up to the level where a cell-killing effect may impart a downward trend; (2) suitable models may differ according to the LET of the radiation; (3) despite the first, there may be tumors whose dependence upon dose is, for practical purposes, best represented by discontinuous, or threshold, models, e.g., skin cancer and bone cancer, and (4) for epidemiologic work, the linear-quadratic is a reasonable function for most sites with or without an exponential term for the cell-killing effect, depending on the volume of data; the breast and the thyroid may be exceptions.

A dose-response model for radiation carcinogenesis is, however, incomplete in the sense that factors other than dose will govern the response, e.g., dose-rate, host characteristics, and temporal patterns descriptive of time to initiation of effect and of duration of effect.

Epidemiologists are looking to the evolution of a more adequate theory of radiation carcinogenesis and to the results of experimental work for guidance as to the choice of dose-response models for human data. For regulatory purposes, however, a choice of model may be governed by a desire to minimize health effects. Hence, among the models having the best theoretical and empirical support, the one that yields the highest risk estimates is often favored. Even more conservative (higher) estimates would flow from supra-linear functions of the type Baum (1973) has suggested, but there is not yet evidence of the existence of sufficiently large sub-populations with higher levels of genetic susceptibility to permit us to rule out the linear hypothesis, and there is no real basis for specifying the parameters of supra-linear functions.

In making empirical risk estimates, and in specifying the gaps in knowledge that limit the validity of those estimates, it has seemed useful to recognize the following determinants of the carcinogenic response to ionizing radiation:

1. Characteristics of the radiation exposure itself
    a. magnitude of dose to target organ or tissue

b. dose-rate
   c. LET
   d. whether other tissues or the whole body are exposed
2. Dose-response relationship
3. Host factors
   a. age
   b. sex
   c. others, to be determined
4. Other environmental factors
5. Differential sensitivity of target organs and tissues to the carcinogenic potency of ionizing radiation
6. Time-response relationships
7. Relationship of the radiogenic excess to the natural incidence
8. Incidence versus mortality

For all these determinants, our knowledge is more or less incomplete, but the fact that so many can be listed is some indication of the breadth of our knowledge. Provisional comments and questions follow.

Dose to target tissue: An organ dose obtained as an average value over the entire organ provides a more reasonable basis for prediction, for comparison of organ-specific effects, etc., than a badge reading or a skin dose, but may be inferior to a measurement that is confined to a particular subset of cells within which the critical change occurs. Also, if an average dose to an organ or type of tissue is relevant, why is not the volume of tissue exposed, or the number of cells exposed, also relevant?

Dose-rate: Although NCRP Report No. 64 has recommended a dose-rate factor that should be applied to low-LET radiation effects when extrapolating from estimates obtained at high doses and high dose-rates (NCRP, 1980), great uncertainty attaches to the proper numerical value suitable for human carcinogenesis, and to whether, indeed, there should be a single value, owing to the dearth of relevant information. For breast cancer, the human data suggest that any dose-rate adjustment may be in the range of 1 to 2 (Land et al., 1980).

LET: Although there is general agreement that high-LET radiation is more carcinogenic, than low-LET radiation, the relation between them has not been satisfactorily quantified and may also vary by tumor site. This relationship, of course, depends partly on the respective dose-response functions appropriate for high- and low-LET radiation.

Whole or partial-body exposure: Comparative information on cancer following whole-body versus partial-body exposure is much too limited to indicate whether, and to what extent, their effects may be quantitatively different. The BEIR III Report (NAS/NRC, 1980) (p. 413) makes the point that localized x-ray therapy, despite a typically high dose, may involve only small and poorly defined fractions of the total skeleton, and that radiogenic bone sarcomas from bone-seeking radionuclides are typically found at skeletal sites most commonly involved in naturally occurring sarcomas. Also, if the carcinogenic force of radiation on an organ depends on a simultaneous effect on another organ, e.g., as might be the case if a hormonal change would affect the outcome, then one might expect whole- and partial-body exposure to differ in some instances, especially at high doses.

Dose-response model: There is incomplete agreement on the most suitable form of dose-response function for low-LET radiation, and there is no present expectation that human observations will prove decisive. Rather, it appears that appropriate models will require a more adequate understanding of radiation carcinogenesis itself.

Host factors: age-at-exposure: The magnitude and temporal distribution of radiogenic cancers clearly depend on age-at-exposure, but in no simple, uniform fashion. The dependence probably varies by tumor site. Of particular interest is the question whether the embryo or fetus is very much more sensitive to the carcinogenic action of ionizing radiation than the infant or the young child. Also of interest is the dependence of the temporal distribution of the age-at-exposure effect on the age-specific pattern of natural incidence.

Host factors: sex: Absolute risk estimates differ markedly as a function of sex for some forms of cancer, e.g., thyroid cancer and leukemia. These differences are less when the risk estimates are expressed in terms of relative risk.

Other host factors: What other host factors may modify the carcinogenic response to ionizing radiation remains unclear, but it seems plain that hormonal factors, at least, play such a role in breast cancer. Immune competence and genetic characteristics, e.g., those governing DNA repair capabilities, are also possibilities.

Environmental factors: Other carcinogens, such as the products of cigarette smoke, are potentially significant but have been little studied for their influence on radiogenic risks. Investigative interest currently centers on whether such factors interact with ionizing radiation to increase (or decrease) its effectiveness. The lung cancer data in relation to smoking are inconclusive, those on the U.S. uranium miners being suggestive of a synergistic effect of smoking combined with exposure to radon daughters, those on the Swedish iron miners being suggestive of additivity, and those on the A-bomb survivors also being suggestive of additivity.

Sensitivity of specific target tissues: We are confronted with the fact that risk estimates per unit of tissue dose (as currently measured) vary widely among specific organs, whether measured in terms of absolute or relative risk, and that these estimates bear no obvious relation to natural incidence. Until we know the basis for these differences, radiation carcinogenesis will not be understood.

Distribution of radiogenic tumors over time: As with tumors of other etiology, radiogenic tumors begin to appear only after a latency period, followed by a prolonged period of active expression whose length, however, is well known only for leukemia and certain bone cancers. For leukemia, and for bone cancer following exposure to radium-224, the radiogenic excess is distributed in time according to a wave function, but for solid tumors generally, it seems to follow the pattern of the baseline, age-specific incidence. For both solid tumors and leukemia the temporal distribution depends intimately upon age-at-exposure, but we have not learned why this is so. Radiogenic cancers of the breast and lung seem not to occur until an age is reached at which these tumors occur naturally.

Relationships of radiogenic tumors to natural incidence: The dependence of the minimal latent period upon age-at-exposure, and thus upon age at risk of cancer, and the apparent invariance of relative risk by sex for thyroid cancer and leukemia, contrast sharply with the marked differences in relative risk estimates for breast cancer between Japanese A-bomb survivors and North American women exposed to diagnostic or therapeutic x rays. The relationship of radiogenic cancer to naturally occurring cancer needs to be better understood if reliable projections of risk beyond the period of actual observations are to be made. The BEIR III Committee (NAS/NRC, 1980) was obliged to make its lifetime projections on the basis of both absolute and relative risk models.

Incidence versus mortality: The bulk of the data that are available on the occurrence of cancer in irradiated populations, as in the general population, pertain to mortality rather than incidence (UNSCEAR, 1977; NAS/NRC, 1980). With few exceptions, therefore, inferences about the effects of radiation on the incidence of cancer are based on mortality data adjusted for the expected effects on survival. With certain radiation-induced cancers, the difference between incidence and mortality is often large, e.g., thyroid cancer. The extent to which survival with radiation-induced cancers of other sites may differ from the survival that is characteristic of the "spontaneously" occurring cancers is uncertain.

6.2 Chemicals

6.2.1 Route of Exposure and Site of Action

As discussed in detail in Section 5, with chemical carcinogens, the route of exposure can be an important determinant of the site of cancer induction, particularly with direct-acting carcinogens which may act at the initial point of contact. For the majority of carcinogens, however, which require metabolic activation, the location in the body of activating enzymes is thought to be the major determinant of the site of carcinogenesis.

The chemicals or chemical processes that are known to cause cancer in humans (Table 5.1) differ with respect to their chief sources of exposures, routes of exposures, and main target organs (Table 6.4). It can be seen from Table 6.4 that the majority of these chemicals have been encountered through occupational exposures generally protracted over many years and poorly characterized. In the majority of such cases, moreover, the primary route of exposure has been by inhalation. The principal target organs, in addition to the respiratory tract, have included the urinary bladder and, to a lesser extent, the bone marrow and gastrointestinal tract. It is noteworthy that whereas the breast is highly susceptible to radiation carcinogenesis, no chemicals other than hormones have been observed to induce cancer in the human breast.

6.2.2 Dose-Response Relationship

Much greater quantitative epidemiological information is available on the carcinogenic effects of ionizing radiation than is available on the effects of any one carcinogenic chemical, with the possible exception of cigarette smoke. One reason for this is that much of the radiation epidemiology has involved acute exposures, where the dosimetry and the duration of exposure are relatively well defined. By contrast, most epidemiological studies on the carcinogenic effects of chemicals have been conducted in industrial or environmental settings where the duration of exposure is generally protracted and the dosimetry poor.

Exposures in industry vary markedly according to job, even within a

TABLE 6.4—*Some chemical carcinogens for humans, according to exposure source, exposure route, target organ, and strength of evidence for carcinogenicity[a]*

Agent	Exposure Source and Route	Target Organ(s)	IARC[b] Rating
alcohol	beverage	mouth, esophagus	1
acrylonitrile	occupational (inhal.)	lung, colon (?)	2A
aflatoxin[c]	food	liver	2A
4-aminobiphenyl	occupational (inhal., ingest.)	bladder	1
arsenic (trivalent)	occupational (inhal., ingest.)	lung	1
arsenic (trivalent)	water (ingest)	skin	1
asbestos	occupational (inhal., ingest.)	lung, G.I. (?)	1
auramine	occupational (inhal., ingest.)	bladder	2B
benzene	occupational (inhal.)	bone marrow	1
benzidine	occupational (inhal., ingest.)	bladder	1
chlornaphazine	therapy	bladder	1
bis(chloromethyl)ether	occupational (inhal.)	lung	1
cadmium	occupational (inhal., ingest.)	lung,	1
cadmium	occupational (inhal., ingest.)	prostate (?)	2B
chlorambucil	therapy	bone marrow	2B
chromium	occupational (inhal., ingest.)	lung	1
cyclophosphamide	therapy	bladder, bone marrow	1
diethylstilbestrol	therapy	vagina	1
ethylene oxide	occupational (inhal.)	lung	2B
isopropyl oil	occupational (inhal.)	paranasal sinus	1
melphalan	therapy	bone marrow	1
mustard gas	occupational (inhal.)	lung	1
2-napthylamine	occupational (inhal., ingest.)	bladder	1
nickel	occupational (inhal., ingest.)	nasal cavity, lung	1
soots, tars, mineral oils	occupational (inhal., ingest., dermal)	skin, lung, bladder, gastrointestinal track	1
vinyl chloride	occupational (inhal.)	liver	1
cigarette smoke	inhalation	lung, bladder	1

[a] modified from IARC, 1982.

[b] IARC Rating: Grade 1 is defined as carcinogenic for humans; Grade 2 is defined as probably carcinogenic for humans on the basis of animal evidence and limited human evidence; the evidence for compounds in Grade 2B is more limited than that for compounds in Grade 2A; also see Table 5.1).

[c] Aflatoxin: It is now widely believed that the evidence for liver cancers in humans from aflatoxin exposure is confounded by the (variable) presence of hepatitis B, a greater risk factor than aflatoxin.

given plant. Also, industrial processes change over the years, and exposure measurements are generally infrequent. Furthermore, job records within a given industrial plant are generally incomplete, and job rotation is common. In addition, access to industrial records can be difficult because of concerns about the potential legal consequences of epidemiologic findings. Where studies have addressed industrial populations exposed to ionizing radiation or radioactive materials, the same sorts of difficulties have usually been encountered.

As a result of these difficulties, there is little detailed information comparable to that discussed in the preceding section on ionizing radiation, with respect to such parameters as dose-response relationships, dose-rate effects, the influence of age at exposure, etc. Also, since chemical carcinogens are highly variable in nature, findings with one substance are not necessarily transferable to another.

The existing data on dose-response relationships are generally so crude that, with few exceptions, little can be said other than that the data are not inconsistent with linearity. The exceptions include the curves for lung cancer associated with cigarette smoking (see Figure 6.1) and lung cancer in coke oven workers (EPA, 1984), which show some upward curvature with dose. However, the effort being put into

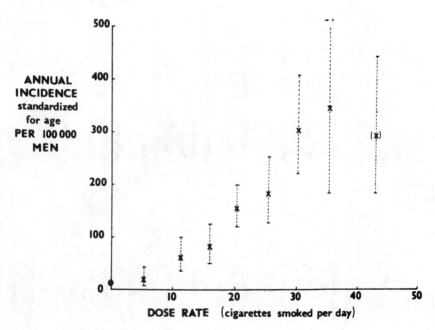

Fig 6.1 Incidence of lung cancer in male cigarette smokers, in relation to the number of cigarettes smoked per day (reproduced from Doll, 1978).

epidemiological studies of environmental and industrial carcinogens has increased rapidly in the last few years.

The few instances of carcinogenesis by radiation and chemicals in which the target organ, duration of exposure, and mode of exposure have been similar are limited chiefly to carcinogenesis in the lung from exposure to arsenic, coke oven emissions, asbestos, cigarette smoke, and radon decay products. Except, perhaps, for cigarette smoke, the dose-response curves are not inconsistent with a linear non-threshold relationship. The dose-response curve for cigarette smoking (Figure 6.1) appears to conform more closely to a power function in which the incidence varies with the dose (number of cigarettes per day) raised to a power of 1.8 (Doll and Peto, 1981).

Although the data are not sufficient to define the shapes of the dose incidence curves in most instances, a markedly increased risk of cancer has been noted in a number of more heavily exposed populations (Table 6.5). In 2-naphthylamine distillers, for example, the latency and incidence of bladder cancer have been observed to vary systematically in relation to the duration of exposure (Figure 6.2). In those with exposures lasting more than five years, the cumulative incidence approached 100 percent (Figures 6.2; 6.3).

TABLE 6.5—*Estimated rates of cancer mortality attributed to various occupational exposures*[a]

Occupation	Cancer Site	Risk (death per million per year)
2-Naphthylamine manufacturing	Bladder	24,000
Nickel workers (employed before 1925)	Lung	15,500
	Nasal sinus	6,600
Mustard gas manufacturing	Bronchus	10,400
Amosite asbestos factory	Lung, pleura	9,200
Rubber mill	Bladder	6,500
Coal carbonizers	Bronchus	2,800[b]
Uranium mining	Lung	1,500
Wood machinists	Nasal cavity	700
Working with cutting oils	Scrotum	400
Printers	Lung, bronchus	200
Shoe manufacturing	Nasal cavity	130

[a] modified from Pochin, 1974.
[b] Includes deaths from bronchitis

With respect to the effects of dose rate and fractionation, there is little evidence on which to compare radiation and chemicals. Although the limited data for radiation imply that dose fractionation has no

72 / 6. CARCINOGENIC EFFECTS OF RADIATION AND CHEMICALS

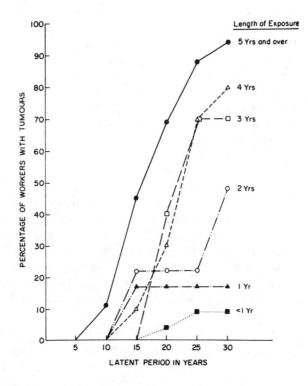

Fig. 6.2 Latency and cumulative incidence of tumors of the urinary bladder in 78 distillers of 2-naphthylamine, in relation to duration of occupational exposure (reproduced from Saffiotti, 1973; based on data from Williams, 1958).

significant effect on the induction of breast cancer in women, as noted above, comparable information for chemical carcinogens does not exist. Nevertheless, there are data suggesting the occurrence of important recovery processes with some chemical carcinogens, e.g., the risk of lung cancer stops rising promptly after discontinuation of cigarette smoking (Doll and Peto, 1981). This plateauing of risk has led some observers to postulate that cigarette smoke acts mainly on late stages in carcinogenesis, perhaps, as a promoting agent, although it also contains agents that can damage DNA (Randerath et al., 1986).

6.2.3 Host Factors

Apart from differences in the age-distribution of chemically-induced cancers, to be discussed below, the limited data on such neoplasmas

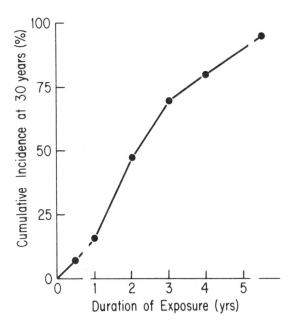

Fig. 6.3 Cumulative incidence of tumors of the urinary bladder 30 years after start of exposure in 78 distillers of 2-naphthylamine and benzidine, in relation to duration of occupational exposure (data from Figure 6.2).

provide no evidence as yet for the influence of host factors on susceptibility such as might be expected from experience with radiation or with carcinogenic effects in laboratory animals.

6.2.4 Distribution of Chemically Induced Cancers in Relation to Time after Exposure

With chemicals, as with radiation, the latency of induced cancers is generally of the order of 10-20 years, except for leukemia, which has a shorter latency (<10 years) after exposure to benzene. For angiosarcoma of the liver, the latency is also somewhat shorter after exposure to vinyl chloride than after exposure to the alpha-irradiation from Thorotrast (Figure 6.4). For lung cancer, there is evidence that the average age at diagnosis in cigarette smokers is relatively constant, regardless of age at the onset of smoking (Doll and Peto, 1981). Arsenic and asbestos are two other chemical carcinogens for which the latency is, likewise, shorter when exposure begins at an older age (Doll and Peto, 1981). In this respect, the data with these chemicals resemble those with radiation.

74 / 6. CARCINOGENIC EFFECTS OF RADIATION AND CHEMICALS

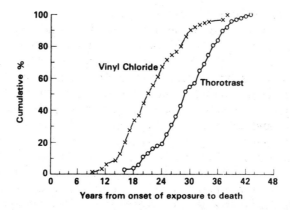

Fig. 6.4 Distribution of death from angiosarcomas of the liver in relation to time after onset of exposure, in persons exposed occupationally to vinyl chloride and persons injected intravenously with Thorotrast (thorium dioxide) (reproduced from Spirtas *et al.*, 1983).

6.2.5 *Relation to Other Environmental Factors and Natural Incidence*

With chemicals, as with radiation, interactions with other environmental risk factors have been observed. Cigarette smoking is a strong enhancing factor for asbestos-induced lung cancer (Selikoff, 1977). Although it is not yet clear whether the relative risk model or the absolute risk model is more appropriate for chemical carcinogenesis, in general, the induction of lung cancer by arsenic is consistent with the absolute risk model.

6.3 Summary

Ionizing radiation and various chemicals have been observed to be carcinogenic for human tissues at high dose levels. In general, whole-body radiation appears to be carcinogenic for many, if not most, tissues of the body, whereas the carcinogenic effects of a given chemical are generally limited to a comparatively small number of target tissues which vary, depending on the chemical in question and the route of exposure.

Sufficient data are available on the carcinogenic effects of radiation at different doses and dose rates (acute as well as chronic exposures) and in males and females of all ages, to enable tentative analysis of the influence of radiological variables and host factors on the dose-incidence relationship. With chemicals, on the other hand, most of the data come from chronic exposures at high dose levels, and the data are too limited, with few exceptions, to enable quantitative analysis of dose-incidence curves. In those cases where data are available, the relationship between incidence and exposure are compatible with non-threshold functions similar to those for radiation-induced cancers.

With chemicals and radiation alike, appearance of the induced cancers is generally preceded by a long latent period, which varies with the type of cancer and age at the time of exposure. The latency for solid cancers is characteristically longer than that for leukemias, and the age-distribution of the induced cancers tends to resemble that of "naturally" occurring cancers of corresponding types. Thus, the excess of cancers resulting from irradiation early in adult life increases with time after irradiation, in parallel with the age-dependent increase in the baseline incidence, such that the radiation-induced excess at any given age is equivalent more nearly to a constant percentage of the corresponding age-dependent baseline incidence than to a constant number of additional cases of cancer.

Whether the same age relationship holds for the effects of carcinogenic chemicals cannot be determined without further data. It is clear, however, that it does not hold for all chemicals, since the risk of lung cancer in ex-cigarette smokers plateaus soon after cessation of smoking. The prompt-plateauing of the risk in ex-smokers implies that cigarette smoke acts largely on late stages of carcinogenesis, as a promoting agent, in this respect differing from radiation.

Further evidence of qualitative differences in modes of action are the data implying multiplicative interactions between the carcinogenic effects of cigarette smoking and those of asbestos. The complexity of such interactions, which are poorly understood as yet, complicates attempts to formulate dose-incidence models applicable to chemical carcinogens, in general. Conversely, the data imply that each exposure situation must be considered individually in assessing the dose-incidence relationship for a given chemical (or combination of chemicals), route of exposure, and population at risk.

7. Extrapolation from Laboratory Models to the Human

7.1 Effects on DNA, Genes, and Chromosomes

Ionizing radiation and genotoxic carcinogenic chemicals cause various molecular changes in DNA, including strand breakage (single-strand and double-strand), base alterations, cross-linkage, and other modifications (Cole et al., 1980, Singer and Grunberger, 1983; Teebor and Frenkel, 1983). Dose-response data imply that such changes can result from the traversal of the cell nucleus by a single ionizing particle (e.g., Cole et al., 1980) or from interactions of DNA with a single electrophilic molecule of carcinogen (Singer and Kusmierek, 1982). The data also imply that the types, frequencies, and repairability of such changes vary, depending on the dose, dose rate, and LET of the radiation (Cole et al., 1980) or the dose and molecular structure of the reactive form of the carcinogen (Singer and Kusmierek, 1982; Singer and Grunberger, 1983).

An exposure to low-LET radiation that is lethal to 50 percent of dividing cells—i.e., approximately 1.5 Sv (150 rem)—produces substantial DNA damage, hundreds of strand breaks per cell. Nevertheless, roughly 50 percent of the cells survive, indicating that much of the damage is reparable (Cole et al., 1980). The effectiveness of repair processes is thus an important factor in determining the fate and effects of lesions in DNA.

In keeping with the production of damage to DNA, the frequency of specific locus mutations in mouse spermatogonia increases as a linear-quadratic function of the dose of low-LET radiation in mammalian diploid cells (NCRP, 1980). The increase per unit dose of low-LET radiation is lower by a factor of about three at low doses and low dose rates than at high doses and high dose rates, corresponding to approximately 6 mutations per 10^6 per locus per 1 Sv (100 rem) (NAS/NRC, 1980; UNSCEAR, 1982). With fast neutrons, the frequency of mutations in such cells increases more steeply, as a linear function of the dose, and is relatively independent of the dose rate. As a result, the RBE of neutrons for mutations in mammalian cells increases to about

30 at low doses and low dose rates (UNSCEAR, 1982). With acute irradiation, the dose-response curves for low-LET and high-LET radiations both reach a maximum in the intermediate dose range and decrease with further increase in the dose, owing at least in part to the killing of cells at risk (UNSCEAR, 1982). On fractionation or protraction of low-LET radiation, the killing of cells is reduced, and the frequency of mutations per unit dose may be higher than that produced by a single brief exposure, depending on the size, number, and timing of exposures (UNSCEAR, 1977; UNSCEAR, 1986).

In cultured human lymphocytes, the frequency of mutations for thioguanine resistance increases as a linear nonthreshold function of the x-ray dose over the range from 10 to 100 mGy (1 to 10 rad) (see Figure 7.1) and is essentially the same whether the dose is delivered in several fractionated exposures or in a single brief exposure (Grosovsky and Little, 1985).

The frequency of interchange chromosomal aberrations in mammalian cells increases as a linear-quadratic function of the dose with low-LET radiation in the low-to-intermediate dose region (NCRP, 1980), the increase approximating 0.1 aberration per cell per 1 Sv (100 rem) in human lymphocytes irradiated *in vitro* (Lloyd and Purrott, 1981). At doses below 50 mGy (5 rad), the frequency of such aberrations per unit dose of low-LET radiation may be somewhat smaller, owing perhaps to more effective repair (Pohl-Rüling, et al., 1983). At intermediate-to-high doses, the frequency per unit dose decreases with decreasing dose rate (Figure 7.2). With high-LET radiation, the frequency of aberrations increases more steeply, as an essentially linear function of the dose, and is relatively independent of the dose rate (Lloyd and Purrott, 1981; UNSCEAR, 1982).

Dose-response relationships for the induction of chromosomal aberrations by chemicals have not been resolved to the same extent as those of ionizing radiation. This is due probably to the greater diversity of types of DNA damage resulting from chemical exposures, the requirement for metabolic activation of many chemicals, the complex and largely unresolved nature of the clastogenic insult, and the magnified effect of the competing interactions between DNA repair and DNA synthesis. While many carcinogens appear capable of inducing chromosomal aberrations (Preston et al., 1983), there are mechanistic differences among them. Clastogenic agents—i.e., chromosomal breaking agents—are generally divided operationally into two classes: S-independent and S-dependent. S-independence denotes the capacity to induce chromosomal aberrations without an intervening round of DNA replication between the insult and the ascertainment. Ionizing radiation and some "radiomimetic" chemicals, e.g., bleomycin, induce chromo-

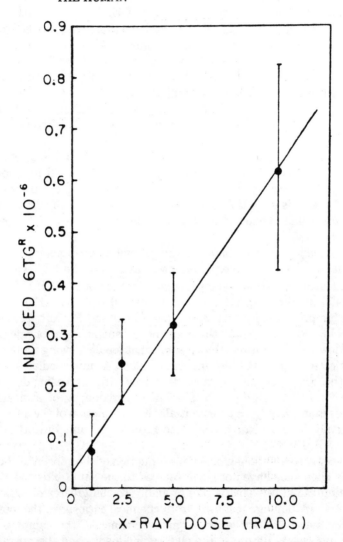

Fig. 7.1 Induction of 6-thioguanine resistance by x rays in TK-6 human diploid lymphoblasts. The line shown, fitted by linear regression analysis, has a slope of 6.0 ± 0.56 TG^R cells per 10^6 Gy (10^8 rad) (from Grosovsky and Little, 1985).

somal aberrations without involving DNA synthesis, while the DNA damage induced by the vast majority of clastogenic chemicals and nonionizing radiation requires DNA synthesis for expression in the form of chromosomal aberrations (Evans and Scott, 1969; Bender et al., 1974). Furthermore, for S-dependent agents, most, if not all, of the resulting chromosomal damage is of the chromatid type, i.e., a single

7.1 EFFECTS ON DNA, GENES, AND CHROMOSOMES / 79

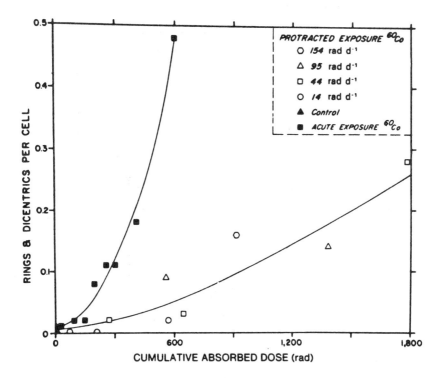

Fig. 7.2 Frequency of rings and dicentrics in Chinese hamster liver cells following acute (50-150 mGy min^{-1}; 5-15 rad min^{-1}) or protracted *in vivo* exposures to ^{60}Co gamma radiation (from Brooks et al., 1971).

chromatid of the metaphase chromosome is involved. S-independent agents induce primarily chromosome-type aberrations during the G1 phase of the cell cycle, primarily chromatid-type during the G2 phase, and a mixture of both types during the S phase (Savage, 1975). These differences in chromosomal aberration production have been interpreted to indicate that the formation of S-independent aberrations involves primarily DNA strand scissions, while the formation of S-dependent aberrations involves primarily DNA base damage. This dichotomy in the induction of chromosomal aberrations, depending on whether DNA replication on a damaged template is required for expression, is exhibited further in the induction of sister chromatid exchanges (SCEs) by these two classes of agents. S-independent clastogenic agents are typically poor inducers of SCEs, while S-dependent clastogenic agents are typically good inducers of SCEs, an observation consistent with the involvement of DNA base damage in the formation of SCEs (Wolff, 1982; Sandberg, 1982; Tice and Hollaender, 1984a,b).

For either class of clastogens, chromosomal aberrations, as indicated earlier, appear to be involved mechanistically in the etiology of experimentally induced tumors in animals and of iatrogenic tumors in humans (Radman et al., 1982; Rowley, 1982, 1984; Yunis, 1983; Sandberg, 1983).

Chemical clastogenesis and mutagenesis both involve a complex series of processes, including pharmacokinetic mechanisms (uptake, transport, diffusion, excretion), metabolic activation and inactivation, production of DNA lesions and their incomplete repair or misrepair, and steps leading to the subsequent expression of mutations in surviving cells or individuals (Table 7.1). Each of the steps in these processes might conceivably involve first order kinetics at low doses (e.g., diffusion, Michaelis-Menten enzyme kinetics) and hence be linear. In principle, therefore, the overall process also might be linear and without threshold.

TABLE 7.1—*Steps involved in chemical mutagenesis*[a]

Steps and processes	Designation	Units (examples)
Exposure to premutagen uptake	Exposure level	ppm; mg; mM
Absorbed premutagen activation deactivation transport	Pharmacologic dose	mg/kg
Mutagen in tissues transport diffusion	Tissue dose	mM; mMh
Mutagen in vicinity of DNA alkylation intercalation	Target dose	mM; mMh
Adducts in DNA repair fixation	Molecular dose	adduct/nucleotide
Mutation	Mutagenic responses	mutations/locus/ gamete mutations/locus/ survivor

From Ehling et al., 1983; modified from Calleman, 1982.[a]

At the same time, non-linear mechanisms may also possibly be involved in the transport, metabolism, repair and elimination processes involved in mutagenesis (Hoel et al., 1983). A single threshold step in such a sequence would suffice to give the overall process a threshold. Furthermore, even if mutagenesis at low doses involved a combination of linear processes, the slope these gave to the dose-response relationship could be orders of magnitude shallower than the corresponding slopes at higher doses. Hence the resulting dose-response curve could appear to have a threshold or quasi-threshold (Hoel *et al.*, 1983).

Among various physiological and biochemical processes in which membranes, enzymes, or organs are known to achieve essentially complete partitioning of chemical species, a familiar example is the excretion of glucose by the mammalian kidney, i.e., at normal blood levels of glucose, the sugar is almost entirely reabsorbed by kidney tubules and essentially none appears in the urine. As the blood glucose level rises, however, the capacity of the tubules is eventually exceeded, and glucose spills into the urine. For certain chemical mutagens, it is conceivable that low levels of the mutagen might circulate in the blood under some circumstances without any effective amount of mutagen reaching the cell nucleus because of the action of various barriers, including scavengers in the blood and cytoplasm.

Once a mutagen reaches the DNA, however, the resulting dose-response relationship should be determined by the nature of the process. A one hit-response will result when a single molecule is sufficient to cause a DNA lesion that eventually gives rise to an induced mutation, e.g., formation of an adduct leading to an altered base, a cross-link, or an intercalation leading to a frame shift. In principle, however, when two independent lesions are required, such as the double breaks needed for a chromosomal rearrangement, the response will be of the two-hit type and will increase as the square of the dose. A mixture of one-hit and two-hit mechanisms is linear at low doses and quadratic at intermediate-to-high doses, resembling the response to x rays (Ehling *et al.*, 1983).

Another factor that can distort dose-response relationships is heterogeneity of dose or response within a population of cells, as may result from differences among cells in degree of differentiation or stage in the cell cycle. Since fixation of mutations and error-prone processes generally involve DNA replication, the farther a cell is from the S-phase, the more likely it is to repair lesions in its DNA before incurring a mutation. Sensitivity to cell killing also is likely to depend on position within the cell cycle, and this may be independent of mutagenesis. Cell-cycle-dependent mutation-induction and cell cycle-dependent killings are both compounded by the likelihood that the mutagen itself will interfere with progression of cells through their cycles as the dose is increased.

In view of the complexity of the many processes involved in chemical mutagenesis, it is not astonishing that a variety of dose-response curves have been reported for mammalian cells exposed *in vitro*. These include both linear and quasi-threshold types of responses (Table 7.2). Whether any of the responses truly have a threshold cannot be determined from the available data.

Observations on the frequency of somatic mutations in the lymphocytes of persons exposed occupationally to chemicals in the workplace are in keeping with the mutagenic effects of chemicals *in vitro* (Tice, 1984).

Another variable complicating the analysis of dose-response relations in chemical mutagenesis is variation in potency. Different chemicals can vary from one another in mutagenic potency by factors as large as 10^6 (Figure 7.3). Furthermore, the magnitude of the variation may vary among different types of cells and different indices of mutagenicity (Tables 7.3, 7.4 and Figure 7.3). Although, in general, there is a correlation between levels of potency in different types of cells, the correlation is not absolute, indicating variations among cells, owing to differences in pharmacokinetics, chromatin structure, DNA repair processes, or other variables. A number of chemical carcinogens, moreover, possess no demonstrable genotoxicity (Upton et al., 1984).

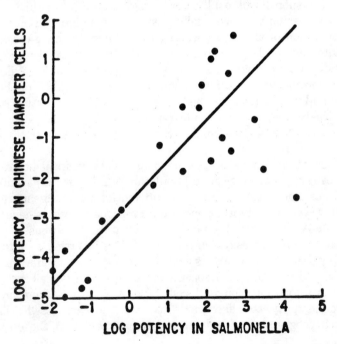

Fig. 7.3 Comparison of mutagenic potencies of chemicals tested in *Salmonella*/microsome and V79 Chinese hamster systems. Ordinate: $-\log P'$, Chinese hamster. Abscissa: $\log P$, Salmonella-liver microsome test. Correlation, $R = 0.74$ (reproduced from NAS/NRC, 1982).

7.1 EFFECTS ON DNA, GENES, AND CHROMOSOMES / 83

TABLE 7.2—*Characterization of dose-response curves for chemical mutagenesis in cultured mammalian cells*[a]

Biological system	Agent[b]	Threshold[c]	Curve shape[d]	Reference
Chinese hamster, V-79 6-TG[R]	MMS, MNU	+	linear	Jenssen, 1981
	EMS,ENU	−	linear	Jenssen and Ramel, 1980
	EMS	−	linear	Van Zeeland, 1978
	BCNU, CCNU, CHLZ CNU, FCNU, *MNU, STRZ*	−	convex downward	Bradley et al., 1980
variant VR-43, ou[R] and 6-TG[R]	MNNG	−	linear	Friedman and Huberman, 1980
Chinese hamster, V79-4B₁, 6-TG[R]	EMS	+	linear	Shaw and Hsie, 1978
CHO, 6-TG[R]	MNU, ENU	−	linear	Thielmann et al., 1979
	EMS	−	linear	O'Neill and Hsie, 1979
	DMS, DES, IPMS, MMS	−	linear	Couch et al., 1978
	ENU	−	linear	O'Neill and Hsie, 1979
	BrdU	−	linear	Irr and Snee, 1979
	ENU	−	linear	Goth-Goldstein and Burki, 1980
CHO, 8-AG[R], SCE proflavine	EMS, ENU, MMS	−	linear	Carrano et al., 1978
CHO, 8-AG[R], 6-TG[R] (repair deficient)	7-BrMeBa	−	linear	Thompson et al., 1982
(Wild type)	7-BrMeBa	±	curvilinear	Thompson et al., 1982
Syrian golden hamster embryo, 8-AG[R]	AF-2	−	linear	Inui et al., 1979
Mouse L5178Y, TK	acridine orange	−	linear	Amacher et al., 1979a
	MMS	−	linear	Amacher et al., 1979a
	MNNG	−	linear	Amacher et al., 1979b
Mouse L5178Y, 6-TG[R]	MMS	−	linear	Cole and Arlett, 1978
TdR	MMS	±	convex downward	Cole and Arlett, 1978
Human fibroblasts, 6-TG[R]	NA-AAF	−	linear at low concentrations	Huang and Lieberman, 1978
	MNNG	±	quadratic	Jacobs and DeMars, 1978
Human fibroblasts, 8-AG[R]				
Human diploid lymphoblasts, 6-TG[R], MIT-2	Aflatoxin, ICR-191	−	convex upward	Thilly 1979, Thilly et al., 1980
	BMS	−	linear	Thilly et al., 1980
GM-130	BMS	+	non-responsive	Thilly et al., 1980
MIT-2	BPL	−	linear	Penman et al., 1979
	MNUT	+	convex downward	Penman et al., 1979

[a] From Ehling et al., 1983.
[b] MMS, methylmethanesulfonate; MNU, methylnitrosourea; EMS, ethyl methanesulfonate; ENU, ethyl nitrosourea; BCNU, bischloroethyl-N-nitrosourea; CCNU, cyclohexylchloroethylnitrosourea; CHLZ, bischloroethyl-2- naphthylamine; CNU, chloroethylnitrosourea; FCNU, cyclohexylchloroethyl-nitrosourea; STRZ, deoxymethylnitrosoureido-D-glucopyranose; MNNG, N-methyl-N-nitro-N'-nitrosoguanidine; DMS, dimethylsulfate; DES, diethylsulfate; isopropylmethanesulfonate; ENU, ethylnitrosourea; BrdU, bromodeoxyuridine; MMC, mitomycin C; 7-BrMeBA, 7-bromoethylbenz[a]anthreacene; NA-AAF, N-acetoxy-2-acetylaminofluorene; BMS, busulfan; BPL, betapropiolactone; MNUT, N-methyl-N-nitrosourethane.
[c] Suggested presence or absence of threshold denoted by following symbols: +, threshold suggested; −, absence of threshold suggested; ±, possibility of threshold suggested.
[d] Shape of dose-response curve.

7. EXTRAPOLATION FROM LABORATORY MODELS TO THE HUMAN

TABLE 7.3—*Mutagenic potency of various chemicals in V79 Chinese hamster cells*[a]

Chemical	Potency[b]
Group 1	*Range $10^{-5}-10^{-4}$*
1-Nitrosopyrolidine	1.0×10^{-5}
N-Nitrosomorpholine	1.5×10^{-5}
Dipropylnitrosamine	2.4×10^{-5}
Diethylnitrosamine	4.8×10^{-5}
1-Methyl-4-nitrosopiperazine	7.7×10^{-5}
3-β-D-Glucopyranosyl-1-methyl-1-nitrosourea	9.0×10^{-5}
Group 2	*Range $10^{-4}-10^{-3}$*
Dimethylnitrosamine	1.3×10^{-4}
Methylpropylnitrosamine	1.4×10^{-4}
Methylazoxymethanol acetate	4.5×10^{-4}
Ethyl nethanesulfonate	7.1×10^{-4}
Dipentylnitrosamine	7.7×10^{-4}
Group 3	*Range $10^{-3}-10^{-2}$*
N'-(Trichloromethylthio)-4-cyclohexane	1.0×10^{-3}
Dibutylnitrosamine	1.0×10^{-3}
2-Deoxy-2-(3-methyl-3-nitrosoureido-)-D-glucopyranose	1.3×10^{-3}
Methyl methanesulfonate	1.5×10^{-3}
Furylfuramide	3.1×10^{-3}
cis-Benzo[a]pyrene-4,5-dihydrodiol	5.8×10^{-3}
Methylnitrosourea	8.4×10^{-3}
Benz[a]anthracene-8,9-diol-10,11-epoxide	9.3×10^{-3}
Group 4	*Range $10^{-2}-10^{-1}$*
Streptozotocin tetraacetate	1.5×10^{-2}
7-Methylbenz[a]anthracene	1.5×10^{-2}
5-Hydroxydibenz[a,h]anthracene	1.6×10^{-2}
1-Hydroxybenzo[a]pyrene	1.6×10^{-2}
Benzo[a]pyrene-3,6-quinone	2.0×10^{-2}
6-Hydroxybenzo[a]pryene	2.2×10^{-2}
3-Hydroxybenzo[a]pyrene	2.4×10^{-2}
Benzo[a]pyrene-11,12-epoxide	2.6×10^{-2}
1-Hydroxy-3-methylcholanthrene	2.8×10^{-2}
p,p'-DDT	3.0×10^{-2}
Benz[a]anthracene-10,11-diol-8,9-epoxide	4.0×10^{-2}
trans-Benzo[a]pryene-11,12-dihydrodiol	4.2×10^{-2}
Benzo[a]pyrene-4,5-epoxide	4.7×10^{-2}
5'-Bromodeoxyuridine	5.6×10^{-2}
Benz[a]anthracene-5,6-dihydroepoxide	6.5×10^{-2}
5-Hydroxydibenz[a,h]anthracene	7.4×10^{-2}
Group 5	*Range $10^{-1}-10^{0}$*
Dibenz[a,c]anthracene	1.0×10^{-1}
3-Methylcholanthrene-11,12-epoxide	2.2×10^{-1}
MNNG	2.9×10^{-1}

7.1 EFFECTS ON DNA, GENES, AND CHROMOSOMES / 85

N-Acetoxy-2-acetylaminofluorene	5.9×10^{-1}
Cytosine arabinoside	7.7×10^{-1}
Group 6	*Range 10^0–10^1*
7,12-Dimethylbenz[a]anthracene	1.6×10^0
3-Methylcholanthrene	2.2×10^0
7-8-Dihydrobenzo[a]pyrene-7,8-epoxide	2.5×10^0
7-Bromomethylbenz[a]anthracene	3.4×10^0
ICR-170	3.8×10^0
[±]-trans-Benzo[a]pyrene-7,8-dihydrodiol-9,10-epoxide	5.0×10^0
7-Methylbenz[a]anthracene-5,6-epoxide	5.0×10^0
1,2-Epoxy-1,2,3,4-tetrahydrobenz[a]anthracene	6.9×10^0
Group 7	*Range 10^1–10^{-2}*
Adriamycin	1.0×10^1
Tetrahydrobenzo[a]pyrene	1.0×10^1
Benzo[a]pyrene	1.4×10^1
Daunorubicin	3.3×10^1
Group 8	*Range 10^2–10^3*
7,8-dihydrodiol-9,10-epoxide	2.0×10^2

[a] Data from Bradley et al., 1980 (from NAS/NRC, 1982).
[b] Mutagenic potency was defined by Bradley et al., 1980 as the concentration (μM) that increases mutation frequency 10-fold above spontaneous background, calculated over a certain dose range and assuming a linear dose-response relationship. These data are expressed as reciprocals to give an ascending scale, in accordance with model groupings.

TABLE 7.4—*Mutagenic potency of various chemicals in the Salmonella/microsome test*[a]

Chemical	Potency[b]
Group 1	*Range 10^{-3}–10^{-2}*
Ethyl p-toluanesulfonate	5.0×10^{-3}
1,2-Epoxybutane	6.0×10^{-3}
Group 2	*Range 10^{-2}–10^{-1}*
Diethylnitrosamine	1.0×10^{-2}
Benzyl chloride	2.0×10^{-2}
N–Nitrosopyrrolidine	2.0×10^{-2}
Dimethylnitrosamine	2.0×10^{-2}
Dimethylcarbamyl chloride	4.0×10^{-2}
N–Nitrosomorpholine	6.0×10^{-2}
Di-n-propylnitrosamine	8.0×10^{-2}

7. EXTRAPOLATION FROM LABORATORY MODELS TO THE HUMAN

Group 3	Range 10^{-1}–10^0
1,2,7,8-Diepoxyoctane	1.0×10^{-1}
1,2,3,4-Diepoxybutane	1.2×10^{-1}
N-Methyl-4-aminoazobenzene	1.4×10^{-1}
Di-n-butylnitrosamine	1.5×10^{-1}
2-Nitrosonaphthalene	1.6×10^{-1}
Ethyl methanesulfonate	1.6×10^{-1}
5-Nitro-2-furoic acid	2.6×10^{-1}
Isophosphamide	2.6×10^{-1}
4-Aminoazobenzene	2.9×10^{-1}
Melphalan	2.9×10^{-1}
4-Acetylaminofluorane	3.0×10^{-1}
N-Hydroxyl-4-aminoazobenzene	3.5×10^{-1}
Styrene oxide	3.7×10^{-1}
Uracil mustard	4.0×10^{-1}

Group 4	Range 10^0–10^1
1-(2-Hydroxyethyl)-2-methyl-5-nitro-imidazole (metronidazole)	1.1×10^0
N-Nitrosoethylurea	1.1×10^0
Methylbis(2-chloroethyl)amine	1.3×10^0
Cyclophosphamide	1.4×10^0
Benzidine	1.4×10^0
Azobenzene	1.4×10^0
p-Dimethylaminobenzenediazo sodium sulfonate	1.8×10^0
Propylenamine	2.0×10^0
Ethylenamine	2.0×10^0
Aflatoxin B2	2.1×10^0
Chrysene-5,6-oxide	2.2×10^0
1'-Acetoxysafrole	2.4×10^0
Mytanthone methanesulfonate	2.5×10^0
4,4'-methylanabis-2-chloroaniline	2.7×10^0
1,1-Diphanyl-2-propynyl-N-cyclohexyl carbamate	2.8×10^0
1,2-Dimethyl-5-nitroimidazole	3.5×10^0
β-Propiolactone	4.1×10^0
N-Nitrosomethylurea	4.4×10^0
10-Chloromethyl-9-methylanthracene	4.6×10^0
Dibenz[a,h]anthracene-5,6-oxide	5.3×10^0
5-Nitro-2-furamidoxime	5.3×10^0
Chlornaphazin	5.6×10^0
1,3-Propane sulfone	6.6×10^0
1-Phenyl-1-(3,4-xylyl)-2-propynyl-N-cyclohexyl carbamate	7.5×10^0
Diazoacetylglycine hydrazide	7.9×10^0
β-Napthylylamine	8.5×10^0
2-Nitrobiphenyl	8.7×10^0
7,9-Dimethylbenz[c]acridine	9.4×10^0
2,7-Bisacetylaminofluorene	9.5×10^0

Group 5	Range 10^1–10^2
9-Aminoacridine	1.0×10^1
Benz[a]anthracene	1.1×10^1

7.1 EFFECTS ON DNA, GENES, AND CHROMOSOMES / 87

Dibenz[a,h]anthracene	1.1×10^1
4-Nitrobiphenyl	1.1×10^1
o-Aminoazotoluene	1.5×10^1
7-10-Dimethylbenz[c]acridine	1.5×10^1
Diazoacetylglycine ethyl ester	1.7×10^1
2-Nitrofluorene	1.8×10^1
Dibenzo[a,j]acridine	1.8×10^1
4-Amino-trans-stilbene	1.9×10^1
7,12-Dimethylbenz[a]anthracene	1.9×10^1
2,3-Epoxypropionaldehyde (glycidaldehyde)	1.9×10^1
Dibenzo[a,i]pyrene	2.0×10^1
7,8-Dihydrobenzo[a]pyrene	2.0×10^1
7-Bromomethyl-12-methylbenz[a]anthracere	2.0×10^1
β-Naphthylhydroxylamine	2.1×10^1
2,7-Diaminofluorene	2.2×10^1
1-Aminoanthracene	2.2×10^1
4-Dimethylamino-trans-stilbene	2.2×10^1
7-Methylbenz[a]anthracene	2.2×10^1
1-Aminopyrene	2.3×10^1
Captan	2.5×10^1
Diazoacetylglycine amide	2.8×10^1
7-Hydroxymethyl-12-methylbenz[a]anthracene	3.0×10^1
4-Aminobiphenyl	3.1×10^1
10-Bromethylanthracene	3.5×10^1
Chrysene	3.8×10^1
Proflavin	3.8×10^1
PNRG	4.0×10^1
N-Hydroxy-2-acetylaminofluorene	4.8×10^1
BNRG	4.9×10^1
N-Acetoxy-2-acetylaminofluorene	5.0×10^1
10-Chloromethyl-9-chloroanthracene	5.5×10^1
3-Methylcholanthrene	5.8×10^1
Folpet	6.4×10^1
Acridine orange	6.6×10^1
2′,3-Dimethyl-4-aminobiphenyl	7.5×10^1
4-Hydroxyaminoquinoline-1-oxide	7.6×10^1
7-Chloromethyl-12-methylbenz[a]anthracene	8.0×10^1
Ethidium bromide	8.0×10^1
9,10-Dichloromethylanthracene	8.8×10^1
ICR-10	9.0×10^1
Group 6	*Range 10^2–10^1*
2-Acetylaminofluorene	1.08×10^2
Adriamycin	1.08×10^2
3-Hydroxybenzo[a]pyrene	1.11×10^2
Aflatoxin M1	1.12×10^2
Aflatoxin G1	1.16×10^2
Benzo[a]pyrene	1.21×10^2
7-Chloromethylbenz[a]anthracene	1.21×10^2
Sodium axide	1.50×10^2
6-Aminochrysene	1.55×10^2

Dibenz[a,c]anthracene	1.75×10^2
2-Aminofluorene	2.05×10^2
α-Nephthylhydroxylamine	2.29×10^2
1-[(5-Nitrofurfurylidene)-amino]hydantoin	2.30×10^2
ICR-170	2.60×10^2
6-Hydroxymethylbenzo[a]pyrene	2.70×10^2
Benzo[a]pyrene-4,5-oxide	2.95×10^2
ENNG	3.50×10^2
Daunorubicin	3.56×10^2
2-Aminoanthracene	5.10×10^2
ICR-191	5.11×10^2
N-Hydroxy-2-aminofluorene	5.83×10^2
3-Methoxy-4-aminoazobenzene	7.47×10^2
Group 7	*Range 10^3–10^4*
2-Nitrosofluorene	1.04×10^3
NCBG	1.375×10^3
1-(5-Nitro-2-thazolyl)-2-imidazolidinone	1.752×10^3
Streptozotocin	1.949×10^3
Aflatoxicol	2.2×10^3
4-Nitroquinoline-1-oxide	2.906×10^3
Aflatoxin B1	7.057×10^3
Group 8	*Range 10^4–10^5*
Azaserine	12.0×10^4
N-[4-(5-Nitro-2-furyl)-thiazolyl]formanide	16.5×10^4
Cigarette-smoke condensate	18.2×10^4
2-(2-Furyl)-3-(5-nitro-2-furyl-acrylamide)	20.8×10^4

[a] Data from McCann *et al.*, 1975, also NAS/NRC, 1983).

[b] Mutagenic potency mutational was defined by McCann *et al.*, 1975, as the number of mutational revertants per nanomole.

7.2 Transforming Effects on Cells in Culture

7.2.1 *Introduction*

Cell culture systems are potentially useful experimental models for identifying environmental carcinogens and assessing their mechanisms of action (IARC/NCI/EPA, 1985). Short-term, inexpensive cell culture assays for carcinogens are available, and these assays have a high predictive ability for the detection of known carcinogens (Heidelberger *et al.*, 1983). In contrast to several other short-term tests, cell transformation assays for carcinogens are not predicated on a theoretical correlation between carcinogenesis and mutagenesis or other genotoxic endpoints. Rather, the endpoints of these assays are related to the neoplastic transformation of a cell. Certain known human carcinogens, which are not readily detected in other short-term tests have been shown to induce cell transformation, e.g., asbestos (Barrett et al., 1984). In addition, these assays can be used to study the enhancement of cellular transformation by human tumor promoters (Kennedy, 1984). Recent advances in the understanding of the molecular basis of carcinogenesis have involved studies using cell transformation assays (Land et al., 1983b). However, cell transformation assays have certain disadvantages when compared to other short-term tests; these include more-specialized technical expertise and facilities, subjective scoring criteria, and a relatively longer time for the assays (Barrett et al., 1984). In addition, the development of cell transformation assays is not as advanced as other toxicity tests, particularly bacterial mutagenicity assays.

Cell transformation assays measure the induction of phenotypic alterations in cells in culture, which are associated with the neoplastic transformation of the cells. These phenotypic changes include morphological alterations, focus formation on a cell monolayer, growth in semisolid media and alterations in growth and/or differentiation (Barrett et al., 1984; Heidelberger et al., 1983). The neoplastic transformation of cells is a multistep process and cells expressing some of these phenotypic changes exist as intermediate or preneoplastic cells. These cells are not tumorigenic but have an increased propensity to give rise to neoplastic progeny.

7.2.2 Systems to Study Neoplastic Transformation of Cells in Culture

Neoplastic transformation of fibroblastic cells has been reported with Chinese hamster lung cells, rat embryo cells, BHK cells, mouse 3T3 cell lines, a mouse ventral prostate cell line, and the mouse C3H/10T1/2 cell line (Heidelberger et al., 1983; Barrett and Ts'o, 1978). The Syrian hamster embryo cells and the various mouse cell lines have been employed most often in studies employing fibroblastic cells. Neoplastic and preneoplastic transformation of normal human fibroblasts has also been reported in recent years (DeMars and Jackson, 1977, Freedman and Shin, 1977; Kakunaga, 1978; Milo and DiPaolo, 1978; Namba et al., 1978; Borek, 1980; McCormick et al., 1980; Sutherland et al., 1980; Silinskas et al., 1981).

A number of experimental systems utilizing cells from a variety of species and tissues have been employed for studies of cell transformation by chemical carcinogens (Tables 7.5 and 7.6). Because fibroblasts were employed in the initial systems, these cells are the best characterized and have been employed for most of the studies to date (Table 7.5). However, 80 percent of all human cancers are of epithelial origin; therefore, systems employing epithelial cells are important in understanding human malignancies and in risk estimation (Table 7.6). At present, neoplastic transformation of epithelial cells *in vitro* by chemical carcinogens has been reported for cells derived from urinary bladder (Summerhayes, 1979; Summerhayes et al., 1981), epidermis (Colburn et al., 1978; Fusenig et al., 1978; Kulesz-Martin et al., 1981; Ananthaswamy and Kripke, 1981), kidney (Boyland and Hard, 1974), liver (Montesano et al., 1973; Williams et al., 1973; Borenfreund et al., 1975; Katsuta and Takaoka, 1975; Mondal, 1975; Schaeffer and Heintz, 1978), mammary gland (Richards et al., 1980); salivary gland (Wigley, 1979), and trachea (Marchok et al., 1977; Marchok et al., 1978; Steele et al., 1979; Pai et al., 1983; Thomassen et al., 1983). Neoplastic transformation of glial cells has also been reported (Laerum and Rajewsky, 1975; Roscoe and Claisse, 1978).

In addition to these systems, virally infected cultures of rat, mouse, and hamster cells have been employed (Heidelberger et al., 1983; Hollstein et al., 1979; IARC Monograph, 1980). Viral and chemical cocarcinogenesis appears to be a sensitive assay for carcinogens. However, this represents enhancement of viral carcinogenesis, at least in some systems, and the significance of these observations to the mechanism of chemical carcinogenesis is unclear.

7.2.3 Detection of Carcinogens with Cell Transformation Assays

Berwald and Sachs (1965) were the first to quantify cell transformation by scoring the number of morphologically transformed Syrian hamster embryo cells following carcinogen treatment. Later studies demonstrated that the number of morphologically transformed colonies observed is dependent on the dose of the carcinogen employed and the number of cells at risk (DiPaolo et al., 1971a). Statistical analysis of the dose-response curve for chemically induced morphological transformation indicates that it fits a "one hit" model (Gart et al., 1979). Results with this system also demonstrate that this process results from the induction of transformed cells, as opposed to selection of preexisting neoplastic cells (DiPaolo et al., 1971b). Dose-response analysis of the induction of variants of rat tracheal epithelial cells suggests a "one-hit" mechanism for this transformation also (Thomassen et al., 1983), although it should be kept in mind that these results relate only to the first phenotypic change in a multistep process of transformation.

In contrast to the results with the above two cell types, certain aspects of the morphological transformation of C3H 10T1/2 cells complicate quantification of the frequency of transformation of these cells and suggest that this conversion is not a one step process. Haber et al. (1977), Kennedy et al. (1980), and Fernandez et al. (1980) have followed up the original observation by Reznikoff et al. (1973) that the transformation frequency, i.e., number of transformed foci per cell treated after correcting for cell killing, is dependent upon the initial number of cells at risk. Extensive studies by Kennedy et al. (1980) with x-ray-induced transformation have demonstrated that the absolute yield of transformed cells is constant over a wide range of initial cell numbers. This is in contrast to results with Syrian hamster embryo cells (DiPaolo et al., 1971b).

According to Kennedy et al. (1980), these observations suggest that the transformed clones do not occur as the direct consequence of carcinogen treatment. Rather, these authors propose a two-step model to explain the results. The initial change induced by the carcinogen apparently occurs in a large number, perhaps all, of the cells. This change does not result directly in the transformation of the cells but increases the probability of their transformation as a rare, secondary event (Kennedy, 1984).

Heidelberger and co-workers (Fernandez et al., 1980) have made similar observations with C3H 10T1/2 cells and have proposed a "probabilistic theory" to explain the formation of transformed foci following 3-methylcholanthrene treatment of these cells. Their theory is similar to that proposed by Kennedy et al. (1980) in that two steps must occur for cell transformation. The first step is the "activation" of a large percentage of the cells by the carcinogen, which occurs with a probability p_1, and the second step is the transformation of the activated cells, which occurs with a probability p_2 per cell generation. The authors have derived a mathematical equation which predicts the frequency of focus formation based on the probability of these two steps (p_1 and p_2) plus the probability of deactivation per cell generation of the carcinogen-activated cells, which is termed p_3.

Mordan et al. (1982) have suggested that the difficulty in quantifying focus formation in C3H 10T1/2 cells is due to the suppressive effects of normal cells on the expression of focus formation by the transformed cells. These authors suggest that a minimum colony size of $>$ 100 transformed cells at confluence is required for the formation of a transformed focus. The suppression or reversion of morphological transformation of C3H 10T1/2 and Balb/c 3T3 cells has been reported ppeviously (Sivak and Van Duuren, 1967; Brouty-Boyé et al., 1979; and Brouty-Boyé and Gresser, 1981). Haber and Thilly (1978) originally suggested that carcinogens affect at least two parameters in C3H

10T1/2 transformation. The first was the induction of the potential for transformation which occurred in a large percentage of these cells. This is analogous to the activation step proposed by Fernandez et al. (1980). According to Haber and Thilly (1978), this induction occurs in nearly all of the cells and is *not* dose dependent. The second effect of carcinogen treatment suggested by Haber and Thilly (1978) is to influence cell-cell transformed potential of the cells.

At present, as discussed in Section 7.2.6, it is apparent that the expression of morphological transformation of C3H 10T1/2 cells is not a one-step process. The first step appears to be a rapid event (Backer et al., 1982) that occurs in a high percentage of the cells (Haber and Thilly, 1978; Fernandez et al., 1980, Kennedy and Little, 1980). The second step could either be a second qualitative change in the cells that occurs at a low frequency during the growth of the cells or at confluence (Fernandez et al., 1980; Kennedy and Little, 1980; Kennedy et al., 1984; Barrett and Elmore, 1984) or an amplification of the transformed cells to overcome the suppressive effects of the non-transformed cells (Haber and Thilly, 1978; Mordan et al., 1982). Further experiments are needed to elucidate the mechanism of transformation and the relevance to neoplastic progression *in vivo* of carcinogen-induced events in C3H 10T1/2 cells.

It is not known whether the transformation of Balb/c 3T3 cells is also a two-step process, like that in the other subtetraploid murine cell line, C3H 10T1/2. However, the results with the preneoplastic hamster cell line BHK are more consistent with a one-step process (Bouck and DiMayorca, 1976).

7.2.4 Dose-Response Data for Transformation of Cells by Chemicals and Radiation

Few detailed dose-response curves for cell transformation systems have been published; however, two carcinogens, benzo[a]pyrene and x rays, have been studied extensively. Transformation of Syrian hamster embryo cells by polycyclic hydrocarbons does not correlate with the cytotoxicity of the carcinogen treatment (DiPaolo et al., 1971a; Umeda and Iype, 1973). Huberman and Sachs (1966) and DiPaolo et al. (1971a) reported that the logarithm of the frequency of morphological transformation of Syrian hamster embryo cells increased linearly with the logarithm of dose. The slope of this line was approximately unity in both studies, suggesting a one-hit model for this change. Gart et al. (1979) have developed statistical methods to analyze these data and have confirmed that the results are fitted very well by one-hit curves at all but the highest doses. The deviation at high doses is possibly due to cytotoxicity of the chemical. Two-hit and multi-hit models were rejected by their analysis. The one-hit model also holds for the results with x rays plus benzo[a]pyrene-induced transformation.

Borek and Hall (1973) determined the dose-response relationship for the morphological transformation of Syrian hamster embryo cells following x-ray treatment. The logarithm of the transformation frequency per surviving cell increased curvilinearly with the logarithm of dose from 0.01-1.5 Gy (1-150 rad). However, a linear response with a slope of 1 also fits within the standard deviation of the data points, which is consistent with the one-hit model of transformation o'served with benzo[a]pyrene-induced transformation (Huberman and Sachs, 1966; DiPaolo et al., 1971a). Interestingly, the authors were able to detect cell transformation with an x-ray dose of only 0.01 Gy (1 rad). The x-ray dose-response curve reached a plateau from 1.5-3 Gy (150-300 rad) and declined at 6 Gy (600 rad). The surviving fraction with 1.5 and 3 Gy (150 and 300 rad) of x rays was 0.86 and 0.75, respectively. Therefore, the plateau of transformation was not associated with a high level of cell killing. The surviving fraction with 6 Gy (600 rad) of x-ray treatment was only 0.12-0.13, which may account for the decline in the dose response curve (Borek and Hall, 1973). Dose-response curves for transformation for C3H 10T1/2 cells have been reported and show an exponential increase in transformation frequency (foci per surviving cell) similar to that observed with hamster embryo cells over the dose range of 0.1-4 Gy (10-400 rad) (Terzaghi and Little, 1976a; Han and Elkind, 1979). However, in the experiments with mouse C3H 10T1/2 cells a plateau in the transformation frequency at doses of 6-14 Gy (600-1400 rad) was observed (Terzaghi and Little, 1976b; Elkind and Han, 1979) rather than a decline as observed at 6 Gy (600 rad) by Borek and Hall (1973) and in the hamster cells. Also a dose of 0.1 Gy (10 rad) yielded a doubling of the transformation frequency in the hamster cells (Borek and Hall, 1973). The doubling dose for mouse cells was 1 Gy (100 rad) (Terzaghi and Little, 1976b).

The effects of split doses of x rays on the transformation of Syrian hamster embryo cells and C3H 10T1/2 cells have been studied. Borek and Hall (1974) observed that if an x-ray dose of 0.5 or 0.75 Gy (50 or 75 rad) was divided into two fractions separated by 5 hours, more cell transformation was observed than if the same dose was given in a single exposure. Cell killing, on the other hand, decreased with split doses versus single exposures. Similar split-dose studies with C3H 10T1/2 cells have indicated that the observed effect varies with the x-ray dose used. At low doses (< 1 Gy; < 100 rad) an enhancement in cell transformation is observed with two split doses (Miller and Hall, 1978;

Miller *et al.*, 1979); with intermediate doses (1-2 Gy; 100-200 rad) no effect of split doses is observed (Terzaghi and Little, 1976b; Miller and Hall, 1978; Miller *et al.*, 1979); and at high doses (3-8 Gy; 300-800 rad), decreased transformation is observed with split doses (Terzaghi and Little, 1976b; Miller and Hall, 1978; Miller *et al.*, 1979; Han and Elkind, 1979). With protraction, the transforming effectiveness of gamma radiation has generally been observed to diminish, whereas that of high-LET radiation has been enhanced (Hill *et al.*, 1984).

Few split-dose experiments have been performed with chemicals. Popescu *et al.* (1984) observed that with split doses of carcinogen separated by 2 to 24 hours only N-acetoxy-2-fluorenylacetamide enhanced transformation of Syrian hamster embryo cells while doses of N-methyl-N'-nitro-N-nitrosoguanidine, mitomycin C or ultraviolet light were less effective than single doses, and no effect of dose fractionation was observed with methyl methanesulfonate.

7.2.5 *Modifiers of Chemical- and Radiation-Induced Cell Transformation*

The neoplastic transformation of different cells by chemicals or radiation can be either enhanced or inhibited by a variety of agents. Promotion of transformation of Syrian hamster embryo cells and/or C3H 10T1/2 cells by the mouse skin tumor promoter 12-0-tetradecanoyl-phorbol-13-acetate (TPA) has been observed with a number of initiating agents, including x rays, UV, benzo[a]pyrene, 3-methylcholanthrene, carcinogenic metals, aflatoxin B, N-methyl-N'-nitro-N-nitrosoguanidine and formaldehyde (Kennedy, 1984). Other promoting agents, such as epidermal growth factor, cigarette smoke extract, and saccharin, also are active in enhancing transformation by chemicals and/or radiation (Kennedy, 1984).

A number of inhibitors of carcinogen- and radiation-induced cell transformation have been identified. These include protease inhibitors (antipain, leupeptin, soybean tryspin inhibitor), vitamin A analogs (retinoids), vitamin C (ascorbic acid), glucocorticoids, and lymphotoxins (Kennedy, 1984).

7.2.6 *Mechanistic Studies of Cell Transformation*

Neoplastic transformation in cell culture is a multistage process, and the mechanisms of the different steps in this process may vary. For example, some aneuploid cell lines, which are nontumorigenic, have some properties of neoplastic cells and progress to neoplastic cells more readily than normal, diploid cells. Hence, they are therefore, considered preneoplastic cells. The current evidence suggests that transformation of normal, diploid cells to preneoplastic cells may occur by a mechanism different from the mechanism of transformation of preneoplastic cells to neoplastic cells.

The nature of a preneoplastic cell *in vitro* is not defined clearly and, therefore, it is difficult to discriminate between normal and preneoplastic cells. Some aneuploid cell lines have been demonstrated as preneoplastic by their increased propensity to become neoplastic *in vitro* and *in vivo* (Barrett et al., 1984). For this reason, it is assumed that only early-passage, diploid cells are normal and that all aneuploid cell lines are preneoplastic. Whether establishment as a permanent cell liner *per se* is a preneoplastic alteration is unclear, especially if the cells remain diploid, an event which occurs only rarely.

Among the quantitative cell transformation systems that employ normal early-passage, diploid cells are those using Syrian hamster embryo cells, human fibroblasts, rat tracheal epithelial cells and mouse epidermal cells. Syrian hamster embryo (SHE) cells have been studied to the greatest extent. When SHE fibroblasts are treated with N-methyl-N'-nitro-N-nitrosoguanidine (MNNG), a direct acting mutagen and carcinogen, a high frequency ($>1\%$) of morphologically transformed cells is observed seven days later (Barrett and Ts'o, 1978; DiPaolo et al., 1971b). The frequency of cell transformation in these studies is 20-100 times greater than the frequency of gene mutations at either the HRPT or NA^+/K^+ ATPase locus measured concomitantly in the same cells (Barrett and Ts'o, 1978; Huberman et al., 1976). This may indicate that gene mutations are not important mechanistically in this step of cell transformation. However, different genetic loci can have quite different frequencies of induced mutation, depending on the fraction of all changes within those genes that actually result in detectable phenotypic changes. Specifically, in human fibroblasts the maximum reported frequencies of gene mutation at the adenine phosphoribosyl transferase locus (ARPT) is 2×10^{-2} per cell (Steglich and Demars, 1982). This value approximates the high frequencies of carcinogen-induced preneoplastic transformation of SHE cells.

Induction of morphological transformation of SHE cells can occur in the absence of detectable induction of gene mutations measured concomitantly in the same cells (Barrett *et al.*, 1983). Diethylstilbestrol and asbestos are two examples of chemicals that induce cell transformation but not gene mutations (Barrett *et al.*, 1981, 1983). However, a good correlation is observed between induction of chromosomal mutations (numerical and structural changes) and induction of cell transformation (Barrett *et al.*, 1983; Oshimura *et al.*, 1984).

A variety of chemical carcinogens induces anchorage independence of human cells in a dose-dependent manner (Milo and DiPaolo, 1978; Silinkas et al., 1981; Sutherland et al., 1980). The frequency of this change varies greatly, and frequencies as high as 10^{-2} per surviving cell have been reported (Silinkas et al., 1981; Milo et al., 1981). Maher et al. (1982) demonstrated that DNA repair-deficient human cells from xeroderma pigmentosum patients have an increased sensitivity to UV-induced killing, mutagenesis, and transformation to anchorage independence. A ratio of the frequencies of cell transformation to gene mutation of 2.5 was observed in these studies, although in earlier studies by the same investigators a ratio of 22 was reported following propane sultone treatment (Silinskas et al., 1981). These authors conclude that anchorage independence results from mutations induced by DNA replication on a damaged template (Maher et al., 1982).

Few quantitative studies on the mechanism of the induction of early preneoplastic changes have been reported in epithelial cells in culture (Barrett et al., 1984). Treatment of rat tracheal cells with carcinogens results in the induction of cell variants with enhanced growth in culture. Following treatment of rat tracheal cells with MNNG, the occurrence of "enhanced growth" (EG) variants follows a Poisson distribution, a linear increase in the logarithm of transformation frequency with a logarithmic increase in dose is observed, and the slope of this dose-response curve is approximately one. These observations are consistent with a "one-hit" mechanism for this induction. The frequency of MNNG-induced EG variants is as high as 4 percent in the surviving cells, which is similar to the frequency of morphological transformation following a similar treatment of SHE cells (Thomassen et al., 1983).

There are no general conclusions that can be drawn at this time about the cellular process involved in the late stages of cell transformation. Preneoplastic cells progress with time to the neoplastic state. In some cells this progression involves multiple changes while in other cases this transition apparently occurs by a single step, the mechanism of which is consistent with a gene mutation op genomic rearrangement (Barrett et al., 1984). The rate of progression is increased by mutagen treatment in some cells but not in others. Chromosomal stability and ploidy may affect the mechanism of preneoplastic progression in some systems. Cellular heterogeneity between cell lines and even sometimes between different subclones of the same cell line is often observed (Barrett et al., 1984).

7.2.7 Role of Oncogenes in Multistage Carcinogenesis

Viral oncogenes were initially identified as sequences in the genome of oncogenic retroviruses which are essential for the carcinogenicity of these viruses (Weinberg, 1982). The discovery that homologous genes exist in human and animal genomes not previously exposed to these viruses has suggested that these cellular or proto-oncogenes may play a role in non-viral cancers (Weinberg, 1982). Several lines of evidence now exist to support this hypothesis (Weinberg, 1982; Barrett et al., 1984). Recent studies on the transformation of normal and preneoplastic cells by cloned oncogenes have provided additional support for the multistage model of carcinogenesis as well as new insight into the molecular basis of certain steps in this process (Land et al., 1983a, 1983b).

The ability of a given cell to be converted to the neoplastic state by an oncogene depends on the stage of neoplastic development of that cell and the biological properties of the product of the oncogene. For example, some preneoplastic cell lines can be rapidly altered to the neoplastic state by the introduction of the H-ras oncogene into the cell by DNA transfection (Land et al., 1983b; Ruley, 1983; Newbold and Overell, 1983). This oncogene is unable to transform normal cells by itself under similar conditions. This suggests that some initial event must occur in the normal cell to make it susceptible to the action of the H-ras oncogene, indicating at least two steps in the neoplastic process. When two cooperating oncogenes are introduced into normal rat cells, these cells can be rapidly transformed to the neoplastic state (Land et al., 1983b; Ruley, 1983). Certain oncogenes can immortalize normal cells into preneoplastic cell lines (Ruley, 1983; Rassoulzadegan et al., 1983). These same oncogenes, however, are unable to transform preneoplastic cells into neoplastic cells. Thus, two classes of oncogenes appear to exist—one class operates at an early stage in cell transformation and the other at a late stage. The gene products of the oncogenes that act at an early stage share a common property, i.e., association with the nucleus and/or binding to DNA (Bishop, 1983). The exact significance of this relationship is unknown. In contrast, oncogenes that act to transform cells which are already preneoplastic code for proteins that are usually found in the cytoplasm and/or on cellular membranes, and in some cases these are active as protein kinases (Bishop, 1983).

The implications of these findings are important: first, they suggest that at least two distinct molecular changes are needed for neoplastic conversion; second, they provide some intriguing insight into the possible biochemical nature of these events; and third, the genetic basis for these changes can be discerned by studying the mechanisms of oncogene activation. For example, the H-ras oncogene can be activated by a point mutation (Reddy et al., 1982; Tabin et al., 1982; Yuasa et al., 1983) and/or enhanced expression by promoter insertion (Chang et al., 1982). This provides strong experimental evidence for the role of gene mutations and/or gene transitions in carcinogenesis.

7.3 Effects on Laboratory Animals

Many experimental observations, some of which have predicted human cancer data, have been possible because of organ-specific animal models for chemical carcinogenesis. These models, being similar to their human counterparts, strengthen the association of environmental exposures with cancer development, and aid in the extrapolation of carcinogenesis data from animal species to humans. They also provide a valuable resource for studies of pathogenesis, risk-modifying factors, and cancer prevention.

7.3.1 *Carcinogenesis in the Skin*

The induction of skin cancer in rodents is probably the oldest experimental model under continuous study and has provided much of our basic understanding of the biology of chemical carcinogenesis. Tumors may be induced in skin by repeated application of initiating agents or by initiating agents followed by tumor-promoting agents (see Boutwell, 1978; Poland et al., 1982). Studies comparing tumor induction by these two protocols have suggested that the mechanism of cancer formation by each protocol may be different. Repeated applications of an initiator generally yield *de novo* carcinomas or persistent papillomas, while two-stage protocols yield mostly papillomas, many of which regress. Carcinomas may arise from papillomas late in the two-stage process, and this conversion can be accelerated by re-exposure to initiating agents (Hennings et al., 1973). Skin tumorigenesis involving promoters also differs from repeated initiator application with respect to strain variations in susceptibility and in the types of chemicals that modify each process (Yuspa, 1983). Squamous cell carcinomas are the principal chemically-induced skin cancers in rodents. Melanotic tumors can be induced in hamsters and albino guinea pigs by topical application of polycyclic aromatic hydrocarbons (Deila Porta et al., 1956; Pawlowski et al., 1980). Ionizing radiation is a potent carcinogen for rat and mouse skin, although the former has been studied more extensively. Ionizing radiations of many types (e.g., protons, electrons, and alpha particles) are capable of producing cancers in rat skin (Albert *et al.*, 1969). In general, the type of malignant tumor produced depends on the dose of radiation received. Higher doses (25-60 Gy; 2,500-6,000 rad), which produce more skin damage, generally yield squamous cell carcinomas, while lesser doses (10-20 Gy; 1,000-2,000 rad), yield more basal cell and sebaceous tumors (Albert *et al.*, 1969). A similar dose-effect relationship is observed for benign lesions, i.e., lower doses yield more keratoses while higher doses result in more chronic ulcers and cysts. The latter lesions have a significantly higher risk for malignant change (McGregory, 1979). The results are dependent also on the degree of penetration of the radiation (Albert *et al.*, 1969). The multiplicity of cancers in a single animal increases with radiation dose as well. Remarkably, when the x-ray dose-incidence relationships for humans and rats are compared, the curves are superimposable if the time scales are adjusted to allow for the inter-species difference (a ratio of 37:1) in duration of the life span (Albert *et al.*, 1978).

Little is known concerning the mechanisms involved in radiation induced skin cancer. A single dose is sufficient to induce cancers many months later. Fractionating the doses, even over several hours, generally reduces the tumor yield, suggesting that a repair function can efficiently correct some radiation damage (Burns and Vanderlaan, 1977). Presumably, this relates to DNA repair, but this has not been proven. At high doses, such as 23 Gy (2,300 rad), fractionation actually enhances carcinogenesis, which may result from a reduction in the cell killing that is otherwise extensive at this dose level. There have been no reports of a possible promotion phase in radiation carcinogenesis in skin. Many chronic and dynamic changes occur in the epidermis and dermis during the post-irradiation period, but their promoting potential has not been studied. The precise dosimetry available with ionizing radiation, and the facility for localizing the damage, should encourage more research in this field, particularly for events associated with the initiation phase of carcinogenesis.

7.3.2 *Carcinogenesis in the Liver*

Liver cancer can be induced in rats by localized irradiation (UNSCEAR, 1977) or by a number of different chemicals and dosage schedules (Farber, 1984). Initiation-promotion protocols are also effective inducers of liver tumors. The predominant lesions are preneoplastic nodules, but as in the skin, carcinomas arise late in the process. Altered hepatic cell foci can be identified within a few weeks after the start of carcinogen treatment by specific changes in enzyme activity, inability to store iron, resistance to toxicity from other chemicals (due to altered metabolic capabilities), and increased mitotic rate (Pitot et al., 1978; Williams, 1976). The earliest lesions are termed hyperplastic foci and are reversible if treatment is stopped. If the carcinogenic stimulus is continued, they enlarge but remain phenotypically stable. It is from these precursor lesions that most hepatocellular carcinomas ultimately arise.

Unlike human liver cancer, chemically-induced liver tumors in laboratory animals are not usually preceded by cirrhotic changes, although proliferative stimuli, such as partial hepatectomy, liver toxicity or cirrhosis, enhance the sensitivity of the liver to carcinogens (Weisburger et al., 1967). Under such circumstances, tumors develop after exposure to carcinogens that do not normally affect the liver and they develop after a single exposure to agents usually requiring multiple doses. The type of tumor arising in the liver can also be determined by the carcinogen. Vinyl chloride inhalation or Thorotrast administration produces hepatic angiosarcomas, both in laboratory animals and in humans (Popper et al., 1977). Serum levels of alpha-fetoprotein are elevated in rodents and in primates bearing liver tumors induced by a variety of carcinogens, denoting another similarity between the animal model and its human counterpart (Kroes et al., 1975). Alpha-fetoprotein appears within weeks of the start of exposure to certain hepatocarcinogens.

7.3.3 Carcinogenesis in the Respiratory Tract

Experimental models for respiratory carcinogenesis have been developed in mice, rats, hamsters, and dogs (Montesano et al., 1970; Nettesheim and Griesemer, 1978; Wynder and Hecht, 1976; UNSCEAR, 1977; Dagle and Sanders, 1982; Bair, 1986). With the exception of alveolar adenomas, which occur in some strains of mice, bronchogenic carcinomas of the same cell types as are seen in humans develop in these species, but not necessarily in the same relative frequencies. The lesions are commonly preceded by squamous metaplasia and benign papillary lesions. Bronchial washings often yield abnormal cells well before the tumors are evident clinically. Malignant lesions have been produced by internal and external irradiation (UNSCEAR, 1977; Dagle and Sanders, 1982; and Bair 1986) and by topical application, intratracheal instillation, systemic administration and inhalation of polycyclic aromatic hydrocarbons, nitroso carcinogens, and other agents (see for example, Nettesheim and Griesemer, 1978). Combined topical and systemic exposures to benzo[a]pyrene and a nitrosamine, respectively, may be synergistic in producing such tumors. The intratracheal instillation of particulates, especially ferric oxide, or the inhalation of sulfur dioxide enhances the carcinogenicity of benzo[a]pyrene. Asbestos induces lung carcinomas and mesotheliomas by intratracheal instillation, inhalation, and intrapleural administration. Asbestos also enhances benzo[a]pyrene-induced carcincogenesis (Shabad et al., 1974).

7.3.4 Carcinogenesis in the Mammary Gland

Mammary cancer has been induced experimentally in mice, rats, and dogs (Jull, 1976; Welsch and Nagasawa, 1977; UNSCEAR, 1977). In mice and rats such cancer occurs after exposure to carcinogenic aromatic amines, polycyclic aromatic hydrocarbons, N-nitroso compounds, and gamma-, x- and neutron-irradiation. The mammary cancers are induced by carcinogens given topically or systemically. The yield of tumors increases with dose for all agents. The tumors derived from all treatments are adenocarcinomas of ductal or alveolar origin and are usually preceeded by preneoplastic hyperplastic alveolar nodules, ductal hyperplasia, or ductal papillomas (Medina, 1977). Mammary cancer induced in rats by three injections of N-methyl-N-nitrosourea closely mimics its human counterpart in that it metastasizes to bone and produces hypercalcemia (Guillion et al., 1975).

Both the induction and progression of experimental mammary cancer are hormone-dependent (see for example, Shellabarger, 1981). Ovariectomy or hypophysectomy prior to carcinogen administration eliminates the tumor response. Similar ablations at later times markedly reduce tumor yields, increase latency, and/or cause regression of pre-existing mammary tumors (Clifton and Crowley, 1978). Likewise, adrenalectomy or elevations of prolactin or ovarian hormones enhance tumorigenicity of sub-carcinogenic doses of chemicals or radiation (Welsch and Nagasawa, 1977). Results from these model systems have strongly implicated a role for prolactin as well as ovarian hormones in the pathogenesis of human breast cancer. High-lipid diets after carcinogen exposure enhance mammary carcinogenesis in laboratory rats, suggesting that a promotion phase is being affected (Carroll, 1975), possibly by increasing estrogen and prolactin levels (Weisburger et al., 1977).

7.3.5 Carcinogenesis in the Gastrointestinal Tract

Colon cancer can be induced in mice and rats by irradiation (UNSCEAR, 1977) and a number of chemical carcinogens, including aromatic amines, derivatives of the plant extract cycasin (and closely related 1,2-dimethylhydrazine) and the direct-acting carcinogens N-methyl-N-nitrosoureas and methylnitrosoguanidine (Weisburger et al., 1977; Toth, 1975). The first two groups of compounds can be given by repeated oral or subcutaneous routes. They may be metabolized directly by colonic epithelium or metabolically activated by the liver and delivered to the colon through the bile as conjugated derivatives. Bacterial deconjugation of these derivatives generates active metabolites to interact with colonic cells. Germ-free animals are much less sensitive to carcinogenesis by 1,2-dimethylhydrazine. Hydrazine derivatives are present in the human environment as both synthetic and natural products (Toth, 1975). Direct-acting carcinogens can be given by intrarectal instillation and if given in low dose will initiate colonic mucosa which can then be promoted by bile acids to form tumors (Narisawa et al., 1974).

Systemic administration of colonic carcinogens yields the entire spectrum of changes seen in human colon cancer, including abnormal crypt architecture with disrupted proliferation, papillary and adenomatous polyps, carcinoma *in situ*, and adenocarcinoma (Thurnherr et al., 1973). The distal colon is affected with the greatest frequency, and squamous cell cancer of the anus is also produced. Dietary fat enhances the production of colonic tumors (Nigro et al., 1975), perhaps by increasing bile flow (Narisawa et al., 1974).

In humans, pancreatic cancer is largely of ductal origin, and concurrent ductal atypia, papillary hyperlasia, and carcinoma *in situ* are common. Pancreatic cancer can be induced in rats, guinea pigs and hamsters by azaserine and certain nitrosocarcinogens (Saffiotti, 1975). In hamsters, specific nitrosamines given orally or subcutaneously induced the same spectrum of pancreatic lesions as that seen in humans. These animals also develop hemorrhagic ascites, vascular thromboses, fat necrosis, and distant metastases (Takahashi et al., 1977).

7.3.6 Carcinogenesis in Hematologic Organs

Irradiation induces leukemias and lymphomas of varying types and frequencies in laboratory animals, depending on the conditions of irradiation and the susceptibility of the exposed species and strain (Upton, 1967; Kaplan, 1967; UNSCEAR, 1977). Similarly, such neoplasms are induced by various chemicals, e.g., 3-methylcholanthrene (Irino et al., 1968; Akamatsu et al., 1968), estrogens (Gardner and Dougherty, 1944), 7-12-dimethylbenzanthracene (Sydnor et al., 1962; Huggins and

Sugiyama, 1966), 4-nitroquinoline-1-oxide (Nishizuka et al., 1964), N-nitrosobutylurea (Odashima, 1969, 1970; Yokoro et al., 1970), myleran (Upton et al., 1961), nitrogen mustard (Conklin et al., 1963), 1-akayl-1-nitrosourea (Ogiu, 1978), and benzene (Snyder et al., 1980, Maltoni et al., 1983; Cronkite et al., 1984).

In the hematologic tissues, as in other tissues, carcinogenesis is moderated by age, sex, hormonal status, diet, physiological state, the influence of extraneous chemical and physical agents, and other factors (Kirschbaum, 1951; Upton, 1967; Clayson, 1984). Phorbol, for example, enhances the leukemogenic effects of 7,12-dimethylbenzanthracene (Armuth, 1976), and azathiopurine exerts synergistic effects with N-nitrosobutylurea (Imamura et al., 1973), in the induction of lymphoid tumors in mice. Similarly, urethane (Berenblum and Trainin, 1963) and steriods (Upton and Furth, 1954) enhance the induction of lymphomas in irradiated mice, depending on the conditions of exposure.

Retroviruses have been isolated repeatedly from radiation-induced and chemically induced leukemias; however, their role, if any, in the induction of these neoplasms remains uncertain at this time (Kaplan, 1984).

7.3.7 Carcinogenesis in Other Organs

Other organ sites for which animal models exist include the uterine cervix (Park and Koprowska, 1968), endometrium (Baba and Von Haam, 1967), esophagus (Magee et al., 1976), kidney (Magee et al., 1976), and brain (Goth and Rajewsky, 1974). Cancer of the urinary bladder can be readily induced in rodents by a number of agents given systemically, including 2-naphthylamine, a human bladder carcinogen (Hicks et al., 1978). Intravesical instillation of nitrosomethylurea also produces bladder tumors at high doses and acts as an initiator of carcinogenesis in the bladder epithelium at low doses (Hicks et al., 1978). In the latter case, the artificial sweetners, saccharin and cyclamate, given orally are effective promoters of bladder cancer, while the bladder irritant cyclophosphamide is not. Prostate cancer is the one major human cancer for which no comparable animal model exists; however, prostate cancer occurs spontaneously in aged rats, and can be accelerated in development by administration of androgen (Shain et al., 1979).

7.3.8 Transplacental Carcinogenesis

Certain chemical carcinogens, given to pregnant animals, are potent as inducers of cancer in the offspring (Tomatis and Mohr, 1973). Nitrosoureas are inducers of neurogenic tumors when given transplacentally to rats. Polycyclic aromatic hydrocarbons produce hepatic tumors by transplacental exposure but rarely do so when given to adult mice. In rodents, exposure to carcinogens must occur after the mid-gestation period (around 12 days) for carcinogenicity; prior exposure leads to embryotoxic or teratogenic effects. In primates, it appears that even earlier gestational exposure leads to cancer in the offspring. Tumors in rodents continue to develop throughout the lifetime of the exposed offspring but maintain the transplacental distribution of types of tumors. Initiation of carcinogenesis can be achieved also by transplacental exposure to carcinogens, leading to offspring that develop tumors only if exposed postnatally to promoting stimuli (Goerttler and Laehrke, 1976). Transplacental carcinogen exposure also may yield offspring whose progeny appear to have a higher risk of spontaneous tumor development (Tomatis et al., 1975). Transplacental or neonatal administration of diethylstilbestrol to mice results in vaginal cancer in the offspring (Bern et al., 1976), an observation which was first made well before the association was noted in humans. Prenatal irradiation also may be carcinogenic under certain conditions (Sasaki et al., 1978).

7.4 Summary

The comparative carcinogenic effects of chemicals and radiation in humans are mirrored to varying degrees by similar effects in laboratory animals. In general, moreover, those chemicals that are complete carcinogens in laboratory animals resemble radiation in their ability to damage DNA and to induce gene mutations, chromosomal aberrations, and transformation of cells in culture. At the same time, such chemicals vary widely in potency, target organ specificity, and the relative frequencies with which they induce changes at different levels of biological organization. In addition to these differences, it is noteworthy that there are other classes of chemicals, e.g., tumor-initiators, co-carcinogens, and tumor-promoters, which are capable of carcinogenic effects if administered in conjunction with certain carcinogens, but which appear to be incapable of acting alone, affecting all stages of carcinogenesis, enhancing tumor formation in a wide range of tissues and species, or exerting comparable effects at the different levels of biological organization. Hence, although similarities between the effects of some chemicals and those of radiation in certain model systems imply carcinogenic mechanisms in common, extrapolation from such systems to predict risks for humans are fraught with uncertainty in view of the unexplained differences among chemicals in potency, target organ specificity, and apparent mode of action, as well as the unexplained differences among species in metabolism, homeostatic mechanisms, and tissue susceptibility. These uncertainties notwithstanding, the data derived from model systems supply information about the comparative carcinogenic effects of chemicals and radiation which are invaluable in assessing the risks of chemicals for which human data are lacking and which provide important supporting or amplifying evidence when the relevant human data are limited.

8. Risk Assessment

8.1 Elements Involved in Risk Assessment

The elements involved in risk assessment and risk management are depicted schematically in Figure 8.1. The procedure by which carcinogenic risk assessments are generally performed, and the ways in which data of the various types discussed above are utilized in performing such assessments have been reviewed elsewhere (Krewski and Brown, 1981), and are outlined below.

When epidemiological data are available, the issues to be dealt with include: selection of the appropriate study and control populations, evaluation of exposure levels and tissue doses, determination of the reliability of cancer ascertainment, allowance for the latent period and age distribution of cancers, control of biases and confounding factors, fitting of models to the data to characterize the dose-incidence relationship, and derivation of risk estimates with their associated ranges of uncertainty.

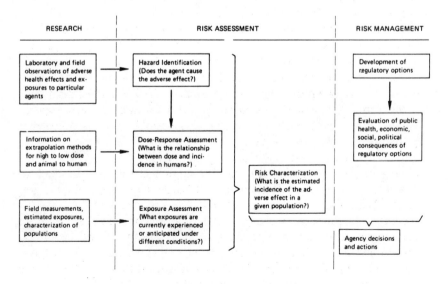

Fig. 8.1 Elements of risk assessment and risk management (from NAS/NRC, 1983).

8.1 ELEMENTS INVOLVED IN RISK ASSESSMENT / 107

A key step in such assessments is to select the anchoring data points, which are observations on cancers induced in man by moderate-to-high-level exposure to the agent in question. These data points indicate more or less directly the risks associated with the corresponding levels of exposure. To assess the risk at lower doses, however, the appropriate dose-response model must be selected for purposes of interpolation (Whittemore, 1978; NAS/NRC, 1980; Krewski and Van Ryzin, 1981). As noted above, this model cannot usually be determined directly from data on human or animal populations, for the reason that sufficiently robust data do not extend to low enough doses. Hence, a dose-response model must be fitted to the data on the basis of expert judgment, taking into account all that is known about the mechanisms of carcinogenesis. An assumption that the risk is proportional to the dose is generally thought to overestimate risks at low doses, as mentioned below (NAS/NRC, 1980; Krewski and Van Ryzin, 1981). Thresholds are not assumed to exist, however, so that some risk is presumed to exist even at the lowest dose levels, although unobserved and unobservable.

For assessing the risks of chemicals, the approach is similar to that used with radiation in those cases where human data are available, but the data are rarely as complete as with radiation. Furthermore, estimation of the dose is usually more difficult with chemicals because of the lack of good monitoring data and other sources of uncertainty (see Section 5). For example, the dose is not usually well quantified even at levels of exposure where carcinogenic effects are conspicuous.

The problem of risk assessment is further complicated by the fact that cancer arises through a multistage process, successive stages of which may be affected differently by different agents, i.e., the data imply that no single process is applicable to all carcinogens, all types of cancer, and all individuals at risk. In any given situation, therefore, the causative role of a particular carcinogen may be difficult to determine.

Most multistage models assume that: (1) a normal cell must undergo two or more stochastic and almost irreversible changes to become transformed into a cancer cell; (2) one or more of the changes may be inherited via the fertilized egg (zygote), and (3) cancer is the consequence of the clonal proliferation of a single cell that has acquired all the necessary changes (Whittemore, 1978; Moolgavkar and Knudson, 1981). According to such a model, any agent that directly or indirectly increases the probability that a cell will undergo one of the changes is a carcinogen, since it increases the likelihood that a cell will ultimately acquire all of the changes necessary for transformation. This type of model also implies that carcinogens affecting different stages in the process can be multiplicative, rather than merely additive, in their combined effects. Also inherent in such models is the concept that the changes necessary for malignant transformation must occur in the proper sequence, some carcinogenic stimuli acting only on early stages and others on later stages, with latent periods for their effects that vary accordingly.

In the absence of definitive human data, risk assessment may have to depend on the results of cancer bioassays in laboratory animals, short-term tests, or other experimental methods. Hence the following issues must be addressed under such circumstances: the ability of the test system to predict risks for man (quantitatively as well as qualitatively); the reproducibility of test results; the influence of species differences in pharmacokinetics, metabolism, homeostasis, repair rates, life span, organ sensitivity, and baseline cancer rates; extrapolation across dose and dose rates, and routes of exposure; the significance of benign tumors; fitting models to the data in order to characterize dose-incidence relationships; and the significance of negative results.

On the basis of present knowledge, the carcinogenicity of an agent for a particular human tissue cannot be predicted accurately by extrapolation from animal data, e.g., a chemical which causes tumors principally in one organ of a given species may cause tumors in other organs, or no tumors at all, in another species. Long-term bioassay results in the mouse have been predictive for the rat in only about 80 percent of cases, and vice versa (Purchase, 1982; Haseman, 1985). The problem is complicated further by the fact that the human population is exposed to myriad agents interacting in various ways, in contrast to animals in the standard bioassay which are ordinarily exposed to one agent at a time.

8.2 Dose-Incidence Relations

8.2.1 *Ionizing Radiation*

The existing evidence does not exclude the existence of a threshold for some (perhaps even all) forms of cancer, but the available epidemiological and laboratory data do not favor such a possibility. Hence, the interpolation models used by national and international experts for estimating the carcinogenic risks of low-level ionizing radiation are generally based on the assumption of a non-threshold dose-incidence relationship (ICRP, 1977; UNSCEAR, 1977; NAS/NRC, 1980; NCRP, 1980; Sinclair, 1981; Rall *et al.*, 1985).

The evidence also indicates, however, that the magnitude of risk per unit dose may vary with the form of cancer, sex, age at irradiation, LET, dose, and dose rate. Moreover, because of the diversity of mechanisms through which irradiation at intermediate-to-high dose levels may be conceived to affect the incidence of different neoplasms, the variety of dose-incidence relationships that has been observed is not unexpected. In view of these differences, it is necessary in risk assessments to consider each type of neoplasm individually, with efforts to integrate insofar as possible all relevant epidemiological and experimental data.

While the relation between incidence and dose may vary from one type of neoplasm to another, the effects of dose rate and LET on the dose-incidence relation generally conform to the patterns illustrated in Figure 3.3. These patterns are consistent with the relationships to be expected if one assumes that carcinogenesis can be initiated in a suitably susceptible individual by a mutation or chromosomal aberration in a single somatic cell. According to this interpretation, the dose-incidence curve for high-LET radiation would be expected to conform, in general, to the following expression:

$$I = C + aDe^{-pD} \tag{8.1}$$

where I is the incidence at dose D, C is the incidence in non-irradiated controls, and a and p are constant coefficients. For low-LET radiation, the dose-incidence curve would conform, in general, to the expression:

$$I = (C + aD + bD^2)e^{-(pD + qD^2)} \tag{8.2}$$

where the symbols are comparable to those above, except for different values of the coefficients a and p plus the addition of constant coefficients b and q (Upton, 1977).

While some of the observed dose-incidence curves conform to the patterns described above, not all of them do, e.g., the curve for breast cancer appears more nearly linear, and the curve for osteosarcomas induced by radium-226 appears more nearly quadratic (NAS/NRC, 1980). Furthermore, the multicausal, multistage nature of carcinogenesis is complex enough so that any simple model is not likely to characterize the dose-incidence relation over a wide range of doses and exposure conditions. At intermediate-to-high doses, for example, radiation can be expected to exert promoting effects as well as initiating effects on tumor formation, through alterations in cell population kinetics, hormonal balance, immunological reactivity, and other changes. At even higher doses, the response can be expected to saturate.

In view of uncertainty about the shape of the dose-incidence curve at low doses and low dose rates, a variety of hypothetical models have been used in an effort to arrive at a range of estimates for purposes of risk assessment. Four of the alternative dose-response models considered by the BEIR Committee (NAS/NRC, 1980) are illustrated in Figure 8.2.

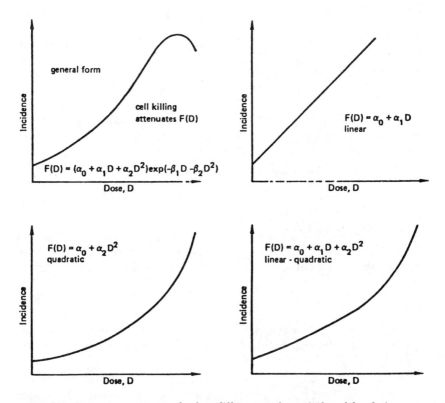

Fig. 8.2 Dose-response curves for four different mathematical models relating cancer incidence to radiation dose (from BEIR III Report, NAS/NRC, 1980).

Whatever mathematical model is assumed for the dose-incidence relationship, it is noteworthy that susceptibility can vary markedly with age, so that the radiation-induced cancer excess at various times after irradiation may more nearly approximate a constant percentage of the natural age-specific incidence than a constant number of additional cases, depending on the neoplasm in question. For some individual neoplasms, but not the leukemias, the data do in fact suggest that the "relative risk" model is more appropriate than the "absolute risk" model (see Section 6.1.7). For all neoplasms combined, also, the excess of radiation-induced cases at different times after irradiation approximates more nearly a constant percentage of the age-specific incidence, than a constant number of additional cases (Doll, 1978; NAS/NRC, 1980; Kato and Schull, 1982; Upton, 1984), conforming more closely to the "relative risk" model (NAS/NRC, 1980).

8.2.2 Chemicals

Evaluation of the carcinogenic risks of chemicals differs from evaluation of the carcinogenic risks of external ionizing radiation in several important respects. First, chemicals vary widely in their capacity to cause cancer ("carcinogenic potency") (Gold et al., 1984), e.g., only a hundred μg/kg (parts per billion) of aflatoxin is associated with an excess of cancer in some populations, whereas a 5 percent content of saccharin in the diet is required to cause an observable carcinogenic effect in a comparable number of animals (Gold et al., 1984). Similarly, dioxin (TCDD), like aflatoxin B1, is 100 million times more potent than saccharin or 1-1-1-tricholoroethane (Crouch and Wilson, 1979; Gold et al., 1984). With some chemicals, no cancers have been observed to be produced in animals at any dose level. As noted above, the variations in potency are attributable in part to pharmacokinetic differences among chemicals (see Section 5.4), which may also influence the uptake, distribution, and retention of some radionuclides (see Section 4.3.3).

Second, the amount of data on individual chemicals is far smaller than the amount of data on radiation. For most chemicals, human data are lacking, and for many chemicals no data at all on carcinogenicity are available (NAS/NRC, 1984).

Third, external penetrating radiation affects cells directly, whereas chemicals and "internally deposited" radionuclides must be transported through the body and, in many cases, metabolized before reaching or acting on the affected cells.

Fourth, chemicals and, indeed, many radionuclides, e.g., iodine-131, often affect primarily only one organ, which may vary among species, whereas external penetrating radiation can affect virtually any organ of the body, depending on the source and energy of the emissions.

Fifth, chemicals often undergo metabolism to yield carcinogenic derivatives, the concentration of which is the relevant dose parameter.

While it is obvious that human data on the carcinogenicity of a chemical should be utilized whenever they are available, no data providing any direct measure of carcinogenic potency for man are available for most chemicals. Although an upper limit to the potency of a chemical can often be inferred from the absence of observed cancers, this limit is usually far too high to be useful in setting exposure standards for public health purposes. For this reason, the carcinogenic risks of a substance can be assessed often only on the basis of its carcinogenicity in laboratory animals or, to a lesser extent, its activity in other bioassay systems.

Criteria to aid in the evaluation of epidemiological and experimental data have been formulated by the International Agency for Research on Cancer (Tomatis et al., 1978; Althouse et al., 1980; IARC, 1982) and the Interagency Regulatory Liaison Group (IRLG, 1979), Office of Science and Technology Policy (OSTP, 1985), and the Environmental Protection Agency (EPA, 1986a, 1986b). These criteria include definitions to describe the adequacy of the data, e.g., the terms "sufficient evidence" and "limited evidence" are defined (Althouse et al., 1980; IARC, 1982). In those situations where "sufficient evidence" for carcinogenicity in man already exists, the human data obviously need to be used in risk assessment. Where there is "sufficient evidence" for the carcinogenicity of a chemical in laboratory animals but not in man, the compound is assumed to present a carcinogenic risk to man (Tomatis et al., 1978; IRLG, 1979; IARC, 1982) but, the magnitude of the risk cannot be estimated with precision, as discussed below. Where no more than "limited evidence" exists, only an upper limit can be placed on the carcinogenic risk that a chemical may pose. While bioassay and short-term "screening" tests may give information on the mode of action of a chemical, such tests are considered to provide only supporting evidence of carcinogenicity and do not provide "sufficient" evidence by themselves.

Estimation of carcinogenic risks on the basis of animal data, however good the data may be, is fraught with difficulty and is inherently imprecise. The considerations reviewed in the previous chapters, although giving some confidence in the inference that a chemical with carcinogenic potency in one species (such as aflatoxin B1) is likely to be potent in another, give no definite prescription for risk assessments. The procedure for making extrapolations from one species to another is often mentioned but rarely discussed in detail. For radiation carcinogenesis, this problem does not arise, since risks can be estimated from human data and the data on animals merely used to help understand the shape of the dose-response curve.

In chemical toxicology, since it has been long known that the acute toxic threshold (LD_{50}) of a chemical is approximately the same in different species if the dose is expressed as a fraction of body weight, this fraction has become the conventional unit in which to express the doses of different chemicals (Dedrick, 1973). The relationship is not precise, however; for some chemicals, the LD_{50} values, even when expressed in these units, can differ by orders of magnitude.

The LD_{50} values for different species are more nearly equal when expressed as a fraction of body surface area. It is customary to retain the old convention for dose units but to correct the LD_{50} values accordingly, i.e., when the LD_{50} dose units are expressed as a fraction of body weight, or mass (m):

$$LD_{50}(man) = (m_{man}/m_{mouse})^{2/3} LD_{50} (mouse) \qquad (8.3)$$

It is natural to consider one or another of these trans-species dose prescriptions for scaling dose-response relationships in carcinogenesis. But in any chronic effect, such as carcinogenesis, another parameter enters; namely, time. Whereas the LD_{50} describes the acutely toxic properties of a chemical, the relevant dose for carcinogenesis is usually accumulated over a long time. One must consider, therefore, the relationship between daily dose, total lifetime dose, and body weight. The difference in life spans between man and mouse—70 years versus 2 years—amounts to a factor 35. Most analyses, however, consider that it is the daily dose that is more relevant, and that the shorter lifetime of the mouse represents the effects of its higher metabolic rate. The difference between these various interspecies dose conversion schemes is illustrated in Table 8.1.

TABLE 8.1—*Scaling factors derived by different methods of adjusting dosage of carcinogen in extrapolating from animals to man*

Basis for species adjustment	Scaling factor(K)[a]	
	K^b human/mouse	K^b human/rat
Daily dose related to body weight[b]	1	1
Daily dose related to surface area[c]	12	6
Daily dose as fraction of diet	4.6	1.8
Lifetime dose	35	35

[a] For explanation of ratios, see Section 8.2.
[b] Modified from Crouch and Wilson (1979). Limits of 0.05 to 20.
[c] Anderson (1983).

Various attempts have been made to determine which scheme, or human/animal scaling ratio (K_{ha}), is correct based on comparative pharmacokinetic data, e.g., Dietz *et al.*, 1983. In an NAS/NRC report on pesticides (1975) the authors assumed that $K_{ha} = 35$, and they studied the limited human data to show that this assumption appeared to be correct within a factor of 100. Crouch and Wilson (1979) and Crouch (1983a, 1983b) have shown that the experimental data for 250 chemicals are consistent with $K_{rat/mouse} = 1$, with a variation in $K_{rat/mouse}$ (between 98th percentiles of the distribution) 1/20 and 20, and that the limited human data are consistent with a variation about $k_{ha} = 1$. Gold *et al.*, (1984) have given values for more chemicals which are also consistent with the above.

Different values have been used by regulatory agencies. The EPA (Anderson, 1983) has used $K_{ha} = (m_h/m_a)^{1/3}$ where m_h and m_a are the mass of the human and the mass of the animal, respectively. The FDA has related the dose in food to the differing fractions of their body weight which different species consume daily. Crouch and Wilson (1981) have argued for using $K_{ha} = 1$ <u>but</u> allowing for variation by taking an upper 98th confidence bound of K_{ha} at approximately 20 and a lower 98th confidence bound of K_{ha} at approximately 0.05.

For interpolating between the lowest dose exhibiting a significantly increased incidence and the baseline (zero dose) incidence, a linear, non-threshold (one-hit) dose-incidence model is often used, although it is recognized that such a model cannot be verified experimentally (Krewski and Van Ryzin, 1981). This model is usually thought likely to overestimate the risk at low doses (e.g., Table 8.2), and so it is often said to estimate the "upper limit" of risk. In addition to the differences in the factors K_{ha} noted above, there may be differences among analysts, e.g., the ratio of the single-hit model calculations of Brown (1977) and those from the FDA is 26, *not* $K_{ha} = 35$ as was assumed in the NAS/NRC, (1975) report on pesticides.

TABLE 8.2—*Estimated human risks from saccharin ingestion of 0.12 g/day*[a]

Method of high-to low-dose extrapolatan	Lifetime cases per million exposed	Cases per 50 million per year
Rat dose adjusted to human dose by surface area rule		
Single-hit model	1,200	840
Multi-stage model (with quadratic term)	5	3.5
Multi-hit model	0.001	0.0007
Mantel-Bryan probit model	450	315
Rat dose adjusted to human dose by milligrams chemical per kilogram body weight per day equivalence		
Single-hit model	210	147
Multi-hit model	0.001	0.0007
Mantel-Bryan probit model	21	14.7
Rat dose adjusted to human dose by milligrams chemical per kilogram body weight per lifetime equivalence		
Single-hit model	5,200	3,640
Multi-hit model	0.001	0.0007
Mantel Bryan probit model	4,200	2,940

[a] From NAS/NRC, 1978

Caution must be observed in such an interpolation, however, and all the relevant data must be used. In at least one case—vinyl chloride— careless use of the procedure could have under-estimated the risk, i.e., in rats exposed to inhaled vinyl chloride monomer, the dose-incidence curve was observed to saturate at the dose at which 20% of rats had angiosarcomas (Maltoni, 1977). Hence, if experimental points at lower doses had not been available, a simple interpolation between the zero dose incidence and the incidence at the maximum dose would have underestimated the computed risk at low doses, as illustrated in Figure 8.3.

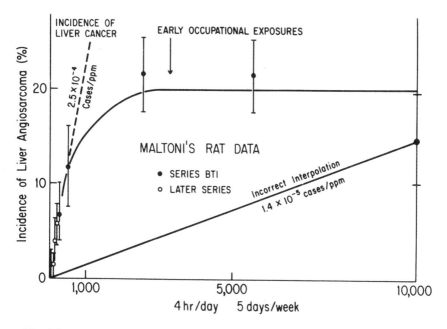

Fig. 8.3 Incidence of liver angiosarcoma in rats versus concentration of vinyl chloride monomer in air inhaled for 4 hours/day, 5 days/week. The line connecting the origin to the highest measured point obviously gives too low a computed risk at the lower dose levels. Also, shown are interpolations indicating the number of cases per ppm inhaled (from Maltoni, 1977).

On the other hand, evidence concerning the different classes of carcinogens (initiators, promoters, co-carcinogens) suggests that a linear non-threshold model may be appropriate only for the initiator class and that a model yielding smaller estimates of the risks at low doses might better represent the dose-incidence relationship for the others. Thresholds might be envisioned to exist for some chemical carcinogens because of their pharmacokinetic properties, i.e., chemicals that must be metabolized before they become carcinogenic, e.g., vinyl chloride, may be handled through non-linear metabolic processes in some instances, leading to thresholds for their carcinogenic effects (Hoel *et al.*, 1983).

At the same time, however, arguments for a threshold, or effective threshold, may be irrelevant for risk analysis. If chemical X, for example, were postulated to have a threshold dose-incidence relationship for stomach cancer, the ingestion of a small amount of substance (X) might be considered to have no effect if stomach cancer were an otherwise rare disease, however, stomach cancer is common enough so that the background level of exposure to chemical X, or some chemical analagous in its carcinogenic behavior, may be conceived to exceed the threshold. Hence the addition of only a small amount of chemical X might be expected to increase the cancer incidence by some finite amount *(I)*, as is illustrated in Figures 8.4a and 8.4b (Peto, 1977; Gaylor and Kodell, 1980; Crump, 1981).

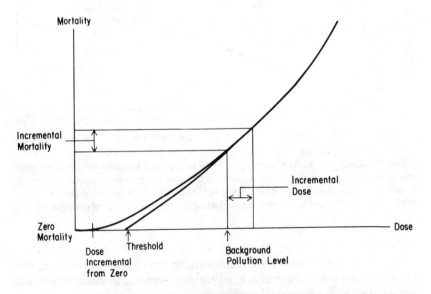

Fig. 8.4a Diagram based on concept of Peto (1977) showing that a small incremental dose of pollutant below a threshold level can produce an increment in mortality in the presence of a background of pollution whereas it may produce no mortality in the absence of a background.

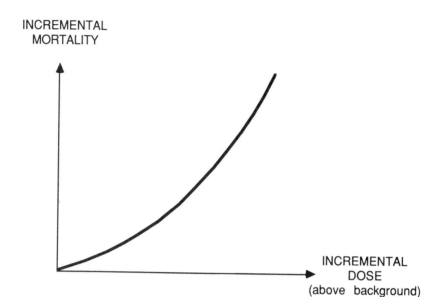

Fig. 8.4b The same graph as in Figure 8.4a, solely for incremental mortality.

When the linear nonthreshold dose-response model is adjusted for background and survival, it becomes:

$$I = 1 - (1-\alpha)\, exp\, \frac{1-\beta d}{1-\alpha} \qquad (8.4)$$

where I denotes the cancer incidence, and α and β are constants. This expression can be derived from a biological one-hit model, and it is often so designated. But this can be confusing, because there are more general derivations. The multistage model without background, but with linearity at each stage, leads to the following formula at low doses:

$$I = \alpha\, d^n\, t^{n-1} \qquad (8.5)$$

where I denotes the incidence, d the daily dose of carcinogen, t the duration of exposure (in days), and n an exponent greater than 1.0. But if we ask, as we always do, whether the events can occur spontaneously—i.e., whether there is a background incidence to which the carcinogen adds an excess—then we find that the excess increases linearly with the dose. This restrictive use of the word multistage can therefore cause confusion.

Other models (Figure 8.5) include the gamma multihit model, which is really only a mathematical version of the simple "multistage" for-

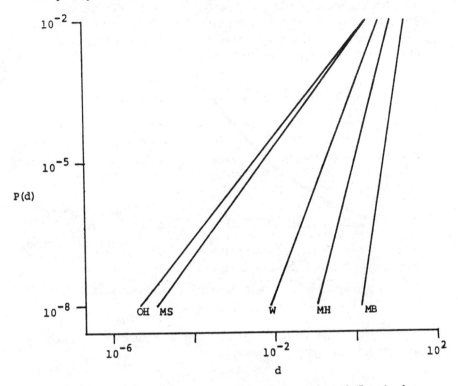

Fig. 8.5 Estimated risk of liver cancer, P(d), in relation to dose of aflatoxin, d, as determined with different dose-incidence models. The models for the different curves are as follows: OH, one-hit model; MS, multi-stage model; W, Weibull model; MH, multihit model; MB, Mantel-Bryan (log-probit model) (from Krewski and Van Ryzin, 1981).

mula but one that allows for a non-integral number of hits or stages (Whittemore, 1978). Whatever its utility as an interpolation model, it is difficult to conceive of one-half a stage in cancer production. If we fit the model, without thought, to the liver angiosarcoma data of Figure 8.2, we find a low dose dependence $(I = \alpha\, d)$, which is surely incorrect.

Another model, widely used in the past, is the Mantel-Bryan probit model (Mantel et al., 1975). This can be derived by assuming that the dose-response relationship for each individual has a threshold, and that the thresholds for different individuals in the population are distributed log-normally. This model gives a lower risk at low doses than does any power law and, therefore, a lower risk than the multistage or proportional models (Figure 8.5). Moreover, when background is included, Crump et al., (1976) and Peto (1977) have shown that it reduces at low doses to the linear-nonthreshold model (Figure 8.4). Therefore, recent risk assessments have tended to include calculations with both a linear nonthreshold model and a threshold model, as two possible extremes, rather than trying to formulate the unknown region with greater mathematical precision than is justified by either the data or our understanding of carcinogenesis.

8.3 Age- and Time-Distribution of Neoplasms

As indicated above, the latent period preceding the appearance of an induced neoplasm may vary with the nature of the carcinogenic stimulus (initiation, promotion, etc.), type of neoplasm, age at exposure, and other factors. These variables must, therefore, be taken into account in estimating the potential impacts of a carcinogen.

The models developed by the BEIR Committee (NAS/NRC, 1980) for assessing the risks of low-level ionizing radiation include consideration of differences in the duration of the latent period for different types of cancer and ages at irradiation, in accordance with the available epidemiological data. Salient features of the models are illustrated in Figure 8.6. Models for assessing the carcinogenic risks of chemicals

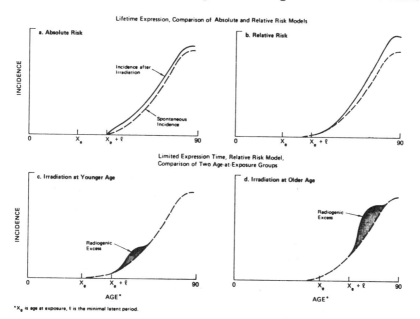

Fig. 8.6 Diagrams illustrating radiation-induced excess of cancer superimposed on baseline age-dependent cancer incidence. Illustrations of possibilities predicted by different models. The patterns shown in panels c and d have been clearly observed thus far only with leukemia and osteosarcoma (from NAS/NRC, 1980).

must also deal with the same variables, taking into account the relevant epidemiological and experimental data.

As is clear from Figure 8.6, the relative risk model assumes that the carcinogenic effects of radiation multiply, rather than merely add to, the carcinogenic effects of other causes responsible for the baseline incidence of cancer at a given age or time following irradiation. This type of interaction may be applicable also in varying degrees to combinations of carcinogenic agents, e.g., cigarette smoking and irradiation, cigarette smoking and exposure to asbestos, etc. Models used in risk assessment should, of course, take such interactions into account when they are known.

8.4 Application to Specific Agents or Exposure Situations

In order to determine the effective dose of a given agent to the population at risk, evaluators need to take into account such variables as the duration and intensity of exposure, the age distribution of exposed persons at the time of their exposure, their sex, state of disease or health, and the estimated concentrations of the agent and its metabolic derivatives in various tissues of the body. Also, insofar as possible, the mode of action of the agent should be characterized in order to enable selection of the appropriate dose-incidence model for use in risk estimation.

In principle, the risks of an agent can be expected to vary, depending on whether it behaves as a complete carcinogen or merely as a promoter or co-carcinogen. Hence the use of *in vitro* mutagenicity tests and bioassays for genotoxicity as potential means of differentiating between these possibilities is being explored. As yet, however, the use of this approach in risk assessment is still in an early stage of development.

8.5 Expression of Risk

Depending on the purposes of a particular risk assessment, the risk may be expressed in different terms. Common measures include the number of additional cases of cancer, the percentage increase in cancer incidence, the number of additional cancer deaths, or the percentage increase in cancer mortality in a population. The loss of life expectancy in the population (in person-years) or the average loss of life expectancy per capita (in minutes, hours, or days) also are helpful measures, because the term "life expectancy" conveys the statistical nature of a risk. The number of working days lost (total per population, or average per capita) is a measure which is useful in reckoning economic costs (lost wages, lost tax revenues, hospital expenses, etc.). Although expression of the risk in such terms can help to put it into perspective with other risks, a full consideration of the risk must also take into account its effects on the overall quality of life.

The acceptability of a risk depends on many factors, of which the magnitude of the risk is only one. Two equal risks may differ in acceptability if one is voluntary, e.g., cigarette smoking or hang-glidding, and the other involuntary, e.g., asbestos exposure in the workplace or being killed by a bus. Also, a large risk for a single individual may be more acceptable to society at large than a smaller risk for many people. Although it is not the purpose of the risk assessment to determine the acceptability of the risk, the assessment should express the risk in ways that enable its implications to be grasped.

Several different expressions of risk may be quoted in an analysis, e.g., a risk averaged over a whole population, an age-specific risk, or a risk for a particularly susceptible individual or subgroup. Such an individual or subgroup might be more susceptible because of genetic predisposition, exposure to another risk factor, e.g., cigarette smoke, or other reasons.

In many cases, although the total quantity of a chemical that is manufactured or ingested is known, the level of exposure of an individual, or a significant subgroup of the population, is not known. This, for example, is the situation in the case of saccharin. The total number of cancers in the population attributable to this chemical, as estimated according to the linear nonthreshold dose-incidence theory (e.g., Table 8.2), is independent of the distribution of saccharin among individuals and represents an "upper limit" estimate of incidence for the total population.

8.6 Uses of Risk Analyses

8.6.1 *Uses for Estimates of Risks of Radiation*

Assessments of the risks of radiation exposure have been made to assist in establishing a basis on which to recommend permissible limits or the appropriateness of already derived limits of radiation exposure for various activities and populations. The permissible limits of exposure vary somewhat for different situations, primarily because of the variation in the risks from causes other than radiation exposure. For example, the recommended maximum annual radiation exposure for an astronaut is 0.5 Sv (50 rem) (other risks are high and the years of potential exposure are limited), for a radiation worker on the ground it is 50 mSv (5 rem) (other risks are lower) while for a member of the general public it is 5 mSv (500 mrem) for occasional exposure and a maximum of 1 mSv (100 mrem) for continuous exposure (other risks are lower yet): However, independent of the magnitude of the other risks, the guiding principle underlying the system of dose limitation is the tenet that the dose should be as low as reasonably achievable (ALARA).

Exposure limits for specific activities have been further limited. For example, Appendix I, Title 10, Part 50, Code of Federal Regulations recommends a design limit of 0.1 mSv y^{-1} (10 mrem y^{-1}) for the maximum calculated radiation dose to any one organ in an individual at the site boundary of a commercial light water nuclear power plant (NRC, 1984).

Medical exposures have been steadily reduced. A chest x-ray in 1945 often was taken on 35-mm film, involving a dose of approximately 10 mGy (1 rad); in contrast, the dose it involves at a good clinic today is about 0.05-0.1 mGy (5-10 mrad) to the thoraxic area. For some diagnostic x-ray procedures, higher exposures are involved—a GI series may involve a dose of 10 mGy (1 rad) to the abdominal area. Table 8.3 lists the effective dose equivalents and the estimated risks associated with various activities.

To a considerable extent, these risk estimates and risk comparisons are merely exposure comparisons. Their interpretation is aided by comparing them with natural background radiation exposure and its variations or comparison to the other risks of a particular activity or to the risks associated with "safe industry".

8.6.2 Uses of Estimates of Risks of Chemicals

It is to be expected that the risks of chemicals will be analyzed for many of the same reasons as are the risks of radiation. Analyses of the risks of chemicals are generally less precise, however, than those of radiation, for the reasons given above.

When exposures to chemical carcinogens have been high enough to cause noticeably large risks, there has usually been little doubt about the need to reduce the risks. It is well known, however, that cigarette smoking — one of the most serious public risks — has not been drastically reduced, e.g., Loeb et al., 1984. In the case of vinyl chloride monomer, which is used to manufacture the millions of tons of plastic in use, no associated toxic effects were known 25 years ago, and no

TABLE 8.3—*Various activities involving radiation exposure, with estimates of the associated risks of cancer mortality*

Activity	Dose (effective dose equivalent)		Fatal cancer Risk per 70 years
One-Time activities			
Planned special exposure: occupational[a]	0.1 Sv	(10 rem)	1.5×10^{-3}
G.I. series[b]	2.45 mSv	(245 mrem)	4×10^{-5}
Chest x-ray[b]	0.05-0.1 mSv	(5-10 mrem)	1×10^{-6}
Repeated activities			
Maximum permissible occupational exposure repeated yearly for 30 years[a]	50 mSv y^{-1}	(5 rem y^{-1})	2×10^{-2}
Lifetime (70 years) exposure to natural background radiation[c,d]	1 mSv y^{-1}	(100 mrem y^{-1})	1×10^{-3}
Power plant boundary[b]	0.1 mSv y^{-1}	(10 mrem y^{-1})	1×10^{-4}
Yearly round trip transcontinental flight[b]	0.06 mSv y^{-1}	(6 mrem y^{-1})	1×10^{-6}
Radon exposure from US homes (average)[d]	2 mSv y^{-1}	(200 mrem y^{-1})	2×10^{-3}

[a]NCRP, 1987b.
[b]UNSCEAR 1977.
[c]Excluding effective dose equivalent from inhaled radon.
[d]NCRP, 1987a.

attempts were made to curtail exposures to this substance. Average exposures in the workplace often exceeded 1,000 ppm, with extremes up to 50,000 ppm. When it was found that vinyl chloride was a carcinogen, and that exposures of long duration led to risks of cancer, industry, labor, and government acted promptly to reduce the exposures. Furthermore, it turned out to be cost-effective to reduce the exposures a thousandfold, to 1 ppm. This level was not chosen by a specific calculation of risk, but by the realization that it was a feasible objective and could be reasonably attained.

Table 8.4 lists estimated occupational risks for various chemicals, as calculated by Albert (1983), in comparison with estimated risks for ionizing radiation. The Table should be used with caution, however, since the calculations for radiation and radon represent "best" estimates, whereas the others are deliberately more "conservative."

TABLE 8.4—*Estimated lifetime cancer risks association with occupational exposure limits for various chemical and physical agents*[a]

Agent	Lifetime Risk of Cancer[b]
Radon (4 WLM y^{-1})	0.03
Radiation (50 mSv y^{-1})	0.02
Vinyl Chloride	0.0007
Actrylonitrile	0.03
Arsenic	0.003
Coal Tar	0.01
Benzene	0.01

[a]From Albert (1983).
[b]Values tabulated denote calculated probabilities of an individual's developing cancer at some time during life following continuous annual occupational exposure for an entire working lifetime at the maximum permissible level.

Other early uses of risk assessment were made by the FDA in 1977 when it considered the risks associated with: (1) the toxic and carcinogenic chemical aflatoxin, which is produced by mold on nut and corn products and, therefore, appears as an unintentional food additive in corn products and peanut butter; (2) the artificial sweetener saccharin, which was found to be carcinogenic in rats; (3) vinyl chloride monomer, which was found to leach in minute quantities from plastic bottles to their liquid contents; and (4) hair dyes, which were found to be mutagenic and carcinogenic in animals. FDA scientists and other experts estimated, with the use of the linear nonthreshold dose-incidence model, the annual numbers of cancer cases that would result from these substances in the U.S. On the basis of these evaluations, the FDA Administrator was urged to regulate the last two items in the series. He could not do so, however, without also considering the regulation of sacaharin and aflatoxin. As is well known, Congress decided not to allow restriction of saccharin, and since it was hard to prevent exposure to aflatoxin, FDA demanded more-rigorous inspection to reduce mold. The regulation of hair dyes and of plastic bottles is still under consideration. Considerations other than risk alone, therefore, decided the issues.

While thousands of workers may be exposed to an agent in the workplace, hundreds of millions may be exposed in the general environment. Consequently, once a linear nonthreshold dose-response relationship is admitted as a possibility, exposures should be limited to a greater degree in the general environment (Table 8.5) than in the workplace. Although natural background radiation and its fluctuations (NAS/NRC, 1980) — or the content of naturally occurring mutagens in foods

(Ames, 1983) — can be used as points of reference, they do not justify exposures from man-made sources.

TABLE 8.5—*Upper-bound risk estimates for suspected carcinogenic air pollutants*[a,b]

Chemical	Upper-bound risk estimates
Acrylonitrile	7×10^{-5}
Allyl chloride	5×10^{-8}
Arsenic	4×10^{-3}
Benzene	7×10^{-6}
Beryllium	6×10^{-4}
Diethylnitrosamine (DEN)	2×10^{-3}
Dimethylnitrosamine (DMN)	5×10^{-3}
Dioxin (2,3,7,8-tetrachloro)[c]	1
Ethylene dibromide	6×10^{-5}
Ethylene dichloride	7×10^{-6}
Ethylene oxide	2×10^{-4}
Formaldehyde	5×10^{-5}
Manganese	4×10^{-4}
Nickel	6×10^{-4}
N-nitroso-N-ethylurea (NEU)	1×10^{-2}
N-nitroso-N-methylurea (NRU)	7×10^{-1}
Perchloroethylene	2×10^{-6}
Trichloroethylene	3×10^{-6}
Vinyl chloride	4×10^{-5}
Vinylidene chloride	4×10^{-5}

[a] From Anderson (1983), based on reports of the U.S. Environmental Protection Agency, Carcinogen Assessment Group 1976-1983.
[b] Excess lifetime risk of cancer for a 70-kg person associated with breathing 1 $\mu g\, m^{-3}$ of the chemical over a 70-year life span.
[c] The potency of dioxin is estimated to be about 1,600 times greater than that of DEN at low exposure levels; therefore, for lifetime exposure to 1 $\mu g\, m^{-3}$, the upper-bound unit risk estimate is 100 percent chance of cancer occurrence. The upper-bound estimate of the potency (slope) for dioxin is 33 $\mu g\, m^{-3}$ or 3.3×10^{-2} $ng\, m^{-3}$.

As risk assessment becomes more sophisticated and is extended to more chemicals, it will also be extended to smaller and smaller risks. Since it is not possible to regulate all risks, an important use of risk analysis must be to decide which chemicals should be regulated and to what degree. Under the Toxic Substances Control Act (TOSCA), all new chemicals must be considered. The prioritization scheme for chemical testing is based upon a set of dichotomous criteria, toxicity, chemical reactivity, etc., which involve qualitative rather than quantitative risk analysis. As procedures for more quantitative analysis become available, the prioritization scheme will become more precise, (NAS/NRC, 1984).

8.6.3 Risk Assessment for Attribution of Risk

In all the uses of risk assessment described above, the goal has been to decide whether a risk is large enough to command attention, whether the risk should be reduced (if possible), or whether the action producing the risk should be deferred, abandoned or modified.

There is another use of risk assessment that is emerging. When an action or exposure affecting the risk of cancer has occurred, how can the subsequent risk of disease be apportioned among its various possible causes? If such apportionment is definite, the word attribution may then be applied to the risks (Rall et al., 1985).

In the simplest of such uses, attribution of risk is performed after a cancer has already occurred in order to assign blame or legal liability. Obvious examples have come from attempts to attribute risks of cancer to previous radiation exposure, but others have also come from previous chemical exposures, e.g., in the case of leukemia following exposure to benzene.

Only in a few instances have the exposures been sufficiently marked to allow attribution, as in the case of a heavily exposed survivor of the A-bombings in Japan who subsequently developed leukemia. From the magnitude of the survivor's radiation exposure and from epidemiological data on the incidence of leukemia in the population at the same exposure level, we might infer that the survivor's increase in risk corresponded to a 2 percent lifetime incidence, in contrast to the normal lifetime leukemia incidence of about 0.5 percent. The attribution would then be 2/2.5, or 80 percent, leaving 20 percent for background radiation and other causes.

If the exposure had been much smaller, the risk calculation would have been less direct and less certain. For purposes of risk reduction in public health, we may choose to err on the pessimistic side in risk estimations. For purposes of attribution, however, we want to make best estimates. Most of the numbers in Table 8.4 are overestimates of the risks. For radiation-induced leukemia, as described in Section 6.1.2, the best dose-incidence model might be linear-quadratic and not linear. Thus, someone exposed to 50 mSv (5 rem) might be considered, on a linear extrapolation basis, to have a radiation related lifetime risk of cancer mortality of 10^{-3} (2×10^{-2} Sv^{-1}; 2×10^{-4} rem^{-1}), or a lifetime risk of mortality from leukemia of approximately 1.5×10^{-4} (0.3×10^{-2} Sv^{-1}; 0.3×10^{-4} rem^{-1}). The "natural" lifetime risk of mortality from leukemia other than chronic lymphocytic leukemia is approximately 56×10^{-4}. Therefore, the percent attribution to radiation according to the linear model would be:

$$\frac{1.5 \times 10^{-4}}{56 \times 10^{-4} + (1.5 \times 10^{-4}} \times 100 = 2.6\% \qquad (8.6)$$

On the basis of a linear-quadratic model for leukemia, however, its attributed risk would be several times smaller. The calculation should also be a function of age at exposure, sex and age at diagnoses (Rall et al., 1985).

For chemicals, the situation is usually even more complex. In the case of someone who has developed angiosarcoma of the liver, however, the tumor is rare enough so that if the person is known to have been exposed to any one of the three known causes, i.e., vinyl chloride monomer, arsenic, or Thorotrast (thorium dioxide), attribution is relatively reliable. In the case of vinyl chloride, the relative risk ratio in exposed workers was 80; hence, the attributed risk in such a worker would be 98–99 percent.

A complication arises, however, when we consider simultaneous exposure to two or more chemicals, e.g., asbestos and cigarettes. The relative risks (RR) of lung cancer in persons exposed to one or both of these agents (Selikoff, 1977) are as follows:

Non-smokers, not exposed to asbestos:	RR = 1
Non-smokers, exposed to asbestos:	RR = 5
Smokers, not exposed to asbestos:	RR = 10
Smokers, exposed to asbestos:	RR = 50

The methods of attribution between asbestos and smoking are obviously more complex and depend on assumptions concerning the interactions between the two agents. Although the field of risk attribution is an old one, its uses for liability are new and fraught with difficulties, some of which are evident from the description above.

8.7 Summary

1. The principles and procedures for assessing the carcinogenic risks of chemicals are similar in essence to those for assessing the carcinogenic risks of ionizing radiation. Risk assessments for the two types of agents differ, however, for the following reasons: (a) chemicals vary in the stage of the carcinogenic process at which they exert their effects, in their uptake and distribution in the body, in their metabolism, toxification, and detoxication, and in their mechanisms of action; (b) for relatively few chemicals are data on carcinogenicity to humans available, and in those instances quantitative information on the dosage of chemical to the target cells is fragmentary; (c) with most chemicals, assessment of the risks for humans must be based on extrapolation from bioassays in laboratory animals or other experimental systems; (d) in few instances has the dose-incidence relationship for chemical carcinogenesis been characterized experimentally over a wide range of

doses and exposure conditions; and (e) marked organ- and species-differences in susceptibility to the carcinogenic effects of chemicals have been observed, the basis for which is still poorly understood.

2. In view of the foregoing, carcinogenic risk assessments for chemicals generally involve greater uncertainty than do carcinogenic risk assessments for radiation.

3. When human data are available for assessing the carcinogenic risks of a chemical, the issues to be considered (as with radiation) include: (1) selection of the appropriate study and control populations, (2) evaluation of exposure levels and tissue doses, (3) determination of the reliability of cancer ascertainment, (4) proper evaluation of the latent period and age-distribution of cancers, (5) control of biases and confounding factors, (6) fitting of models to the data, to characterize the dose-incidence relationship, and (7) derivation of risk estimates, with their associated ranges of uncertainty. For a few chemicals, the existing epidemiological data suffice to define the carcinogenic risks to certain target organs from exposures in the intermediate-to-high dose range.

4. When human data are unavailable or limited, the risk assessment must be based on extrapolation from animal data, short-term tests, and other information. Under these circumstances, the evidence must be weighed with reference to its predictive accuracy, and expert judgment must be exercised in selecting the appropriate model for extrapolation to man.

5. In carcinogenic risk assessments for chemicals, as in those for radiation, risks should be expressed in ways that place them in perspective and make them broadly intelligible. Such an approach will facilitate their interpretation and use.

9. Conclusions

This Report has considered the extent to which the principles and methods that have been developed for use in assessing the carcinogenic risks of ionizing radiation are applicable in assessing the carcinogenic risks of chemicals. Conclusions are as follows:

1. The carcinogenic effects of certain chemicals on man and laboratory animals are similar to those of ionizing radiation. Carcinogenic effects of both types of agents have been noted in virtually every organ of the body, depending on the agent in question, the species, and the conditions of exposure.

2. The cancers induced by both types of agents are individually indistinguishable from those induced by other causes. Hence, their induction can be inferred only on statistical grounds; i.e., from analysis of a dose-dependent increase in their frequency in exposed populations.

3. From the study of human populations exposed to certain chemicals, sufficient data are available to characterize the dose-incidence relationships for some types of cancer at high dose levels; however, as in the case with radiation, the data do not suffice to define the dose-incidence relationship precisely for any form of cancer over a wide range of doses and dose rates. Hence, any carcinogenic risks that may be associated with lower doses of chemicals, such as are of principal concern in public health, can be estimated only by interpolating or extrapolating from effects observed at higher doses and dose rates, based on assumptions about the relevant dose-incidence relationships and mechanisms of carcinogenesis. In the few cases for which incidence data are available over a range of doses, the dose-incidence relationship is not inconsistent with linearity, but this cannot be taken as proof of linearity.

4. Although hundreds of chemicals have been identified as carcinogens for laboratory animals, fewer than 30 are known to cause cancer in man, and for these chemicals, the dose to affected tissues is not known well enough, with few exceptions, to define the dose-incidence relationship except in a general way. In this respect, the carcinogenic effects of most chemicals are far less well known than are those of radiation.

5. Analysis of dose-incidence relationships for chemical carcinogenesis is further complicated by the fact that the dose of a chemical at its biological site of action depends on a number of metabolic and pharmacokinetic factors which can vary with route of exposure, age,

sex, genetic constitution, physiological state, action of other chemicals, and other variables. Such pharmacokinetic variables do not enter into assessment of the risks of radiation, except insofar as they are relevant to the uptake, distribution, and retention of internal emitters. Hence, dosimetry is generally more uncertain with chemicals than with radiation.

6. Chemicals differ widely also in molecular structure, biological activity, and mode of action. Because the relationships among these properties are as yet ill defined and poorly understood, the carcinogenic effects of one chemical cannot be confidently predicted from those of another.

7. The problem is complicated further by the enormous differences in potency among chemicals and in the stages of the carcinogenic process at which they act, e.g., some chemicals affect predominantly late stages, others earlier stages.

8. Another complication arises from the fact that the interactive effects of two or more carcinogens may be mutually additive, synergistic, or antagonistic, depending on the circumstances, and that few such interactions have been well characterized as yet. Furthermore, the combined effects of complex mixtures of chemicals, such as are typically encountered in human life, are virtually unexplored.

9. In spite of the aforementioned differences among carcinogens, the principles for risk assessment that have proven to be useful with ionizing radiation appear to be applicable, within limits, to chemicals, particularly those chemicals that resemble radiation in genotoxicity, cytotoxicity, and in the stages of carcinogenesis affected.

10. In a given exposure situation, however, choice of the appropriate dose-incidence model for risk assessment is a matter of scientific judgment. Such judgment must be based on consideration of all pertinent epidemiological and experimental data, including the results of short-term tests where applicable.

11. Because of the gaps in our present understanding of carcinogenesis and the paucity of human data for most chemicals, risk assessments for chemicals are generally more uncertain than risk assessments for radiation. To improve such assessments, there is particular need for further development and validation of methods for extrapolation from animal data to man and further refinement of methods for evaluating variations in human susceptibility and human exposure, e.g., biological markers.

12. Ultimately, the desired reliability of risk assessment for low-level exposure to chemicals, as well as for low-level exposure to radiation, will require better understanding of the carcinogenic effects of these agents at all stages of carcinogenesis and at all levels of biological organization.

References

AITO, A. (1974). "Different elimination and effect on mixed function oxidase of 20-methylcholanthrene after intragastric and intraperitoneal administration," Res. Commun. Chem. Path. Pharmacol. 9, 701.

AKAMATSU, Y., IKEGAMI, R., WATANABE, K., AND KIKUI, M. (1968). "Amyloidosis in senile C57BL mice by oral feeding of 3-methylcholanthrene in olive oil solution," Gann 59, 489.

ALBERT, R.E., PHILLIPS, M.E., BENNETT, P., BURNS, F., AND HEIMBACH, R. (1969). "The morphology and growth characteristics of radiation-induced epithelial skin tumors in the rat," Cancer Res. 29, 658.

ALBERT, R.E. AND ALTSHULER, B. (1973). "Considerations relating to the formulation of limits for unavoidable population exposures to environmental carcinogens," page 223 in *Radionuclide Carcinogenesis, Proceedings of the 12th Annual Hanford Biology Symposium, May 10–12, 1972,* Sanders, C.L., Ed., A.E.C. Symposium Series, 29 CONF 720505, (National Technical Information Service, Springfield, Virginia).

ALBERT, R.E., BURNS, F.J., AND SHORE, R.E. (1978). "Comparison of the incidence and time patterns of radiation-induced skin cancer in humans and rats," page 49 in *Late Biological Effects of Ionizing Radiation,* IAEA-SM-224/105, (International Atomic Energy Agency, Vienna).

ALBERT, R.E. (1983) "The acceptability of using the cancer risk estimates associated with the radiation protection standard of 5 rems/year as the basis for setting protection standards for chemical carcinogens with special reference to vinyl chloride," Report to Ministry of Labor, Occupational Health and Safety Division, Toronto, Ontario, Canada (Ministry of Labor, Occupational Health and Safety Division, Toronto, Ontario, Canada).

ALBERTINI, R.J. AND ALLEN, E.F. (1981). "Direct mutagenicity testing in man," page 131 in *Health Risk Analysis,* Proceedings of the III Life Science Symposium, Richmond, C.R., Walsh, P.J., and Copenhaver, E.D. Eds. (Franklin Institute Press, Philadelphia).

ALTHOUSE, R., HUFF, J., TOMATIS, L. AND WILBOURN, J. (1980). "An evaluation of chemicals and industrial process associated with cancer in humans based on human and animal data," IARC Monographs Volumes 1 to 20, Report of an IARC Working Group, Cancer Res. 40, 1.

AMACHER, D.E., PAILLET, S., AND RAY, V.A. (1979a). "Point mutations at the thymidine kinase locus in L5178Y mouse lymphoma cells. I. Application to genetic toxicological testing," Mutat. Res. 60, 197.

AMACHER, D.E. PAILLET, S.C., AND TURNER, G.N. (1979b). "Utility of the mouse lymphoma L5178Y/TK assay for the detection of chemical mutagens," in *Banbury Report 2, Mammalian Cell Mutagenesis: The Maturation of Test Systems,* HSIE, A.W., O'NEILL, J.P., AND MCELHENY, V.K., Eds. (Cold Spring Harbor Laboratory Press, Cold Spring Harbor, New York).

AMES, B.N. (1983). "Dietary Carcinogens and Anticarcinogens," Science **221**, 1256.
AMES, B.N., MCCANN, J., AND YAMASAKI, E. (1975). "Methods for detecting carcinogens and mutagens with the *Salmonella*/mammalian - microsome mutagenicity test," Mutat. Res. **31**, 347.
ANANTHASWAMY, H.N. AND KRIPKE, M.L. (1981). "*In vitro* transformation of primary cultures of neonatal Balb/c mouse epidermal cells with ultraviolet-B radiation," Cancer Res. **41**, 2882.
Anderson, E.L. (1983). "Quantitative approaches in use to assess cancer risk," Risk Analysis **3**, 277.
ARMUTH, V. (1976). "Leukaemogenic action of phorbol in intact and thymectomized mice of different strains," Br. J. Cancer **34**, 516.
AUTRUP, H., GRAFSTROM, R.C., BRUGH, M., LECHNER, J.F., HAUGEN, A., TRUMP, B.F., AND HARRIS, C.C. (1982). "Comparison of benzo[a]pyrene metabolism in bronchus, esophagus, colon, and duodenum from the same individual," Cancer Res. **42**, 934.
BABA, N. AND VON HAAM, E. (1967). "Experimental carcinoma of the endometrium," Prog. Exp. Tumor Res. **9**, 192.
BACKER, J.M., BOERZIG, M., AND WEINSTEIN, I.B. (1982). "When do carcinogen treated 10T1/2 cells acquire the commitment to form transformed foci?" Nature **299**, 458.
BAIR, W.J. (1986). "Experimental carcinogenesis in the respiratory tract," page 151 in *Radiation Carcinogenesis*, Upton, A.C., Albert, R.E., Burns, F.J., and Shore, R.E. Eds. (Elsevier Science Publishing Co. Inc., Elmira, New York).
BAIRD, W.M. AND BOUTWELL, R.K. (1974). "Tumor-promoting activity of phorbol and four diesters of phorbol in mouse skin," Cancer Res. **31**, 1074.
BALMAIN, A., RAMSDEN, M., BOWDEN, G., AND SMITH, J. (1984). "Activation of the mouse cellular Harvey-ras gene in chemically induced benign skin papillomas," Nature **307**, 658.
BARRETT, J.C. AND ELMORE, E. (1984). "Comparison of carcinogenesis and mutagenesis of mammalian cells in culture," in *Handbook of Experimental Pharmacology*, ANDREWS, L.S., LORENTZEN, R.J., AND FLAMM, W.D., Eds. (Springer-Verlag, Berlin).
BARRETT, J.C. AND TS'O, P.O.P. (1978). "Mechanistic studies of neoplastic transformation of cells in culture," page 235 in *Polycyclic Hydrocarbons and Cancer*, Gelboin, H. and Ts'o, P.O.P., Eds. (Academic Press, New York).
BARRETT, J.C., WONG, A., AND McLACHLAN, J.A. (1981). "Diethylstilbestrol induces neoplastic transformation of cells in culture without measurable somatic mutation at two loci," Science **212**, 1402.
BARRETT, J.C., GRAY, T.E., MASS, M.J., AND THOMASSEN, D.G. (1982). "A quantitative, clonal assay for carcinogen-induced alterations of respiratory epithelial cells in culture," in *Application of Short-term Bioassays in the Analysis of Complex Environmental Mixtures*, Waters, M. and Sandu, S., Eds. (Plenum Press, New York).
BARRETT, J.C., THOMASSEN, D.G., AND HESTERBERG, T.W. (1983). "Role of gene and chromosomal mutations in cell transformation," New York Acad. of Sci. **407**, 291.

BARRETT, J.C., HESTERBERG, T.W., AND THOMASSEN, D.G. (1984). "Use of cell transformation systems for carcinogenicity testing and mechanistic studies of carcinogenesis," Pharmacol. Rev., 53 S.
BATES, R.R., DEL ANDE EATON, S.A., MORGAN, D.L., AND YUSPA, S.H. (1970). "Replication of DNA after binding of the carcinogen 7-dimethylbenz[a]anthracene," J. Natl. Cancer Inst. 45, 1223.
BAUM, J.W. (1973). "Population heterogeneity hypothesis on radiation induced cancer," Health Phys. 25, 197.
BEEBE, G.W., KATO, H., AND LAND, C.E. (1978a). "Studies of the mortality of A-bomb survivors: 6. Mortality and radiation dose, 1950–1974," Radiat. Res. 75, 138.
BEEBE, G.W., KATO, H., AND LAND, C.E. (1978b). *Life Span Study Report 8: Mortality Experience of Atomic Bomb Survivors, 1950–74*, RERF TR 1–77 (Radiation Effects Research Foundation, Hiroshima, Japan).
BENDER, M.A., GRIGGS, H.G., AND BEDFORD, J. (1974). "Mechanisms of chromosomal aberration production. III. Chemicals and ionizing radiation," Mutat. Res. 23, 197.
BERN, H.A., JONES, L.A., AND MILLS, K.T. (1976). "Use of the neonatal mouse in studying long-term effects of early exposure to hormones and other agents," J. Toxicol. Environ. Health Supp. 1, 103.
BERENBLUM, I. AND TRAININ, N. (1963). "New evidence on the mechanism of radiation leukaemogenesis," page 41, in *Cellular Basis and Aetiology of Late Somatic Effects of Ionizing Radiation*, Harris, R.J.C., Ed. (Academic Press, London).
BERTRAM, J.S. AND HEIDELBERGER, C. (1974). "Cell cyclic dependency of oncogenic transformation induced by N-methyl-N'-nitro-N-nitrosoguanidine in culture," Cancer Res. 34, 526.
BERWALD, Y. AND SACHS, L. (1965). "*In vitro* transformation of normal cells to tumor cells by carcinogenic hydrocarbons," J. Natl. Cancer Inst. 35, 641.
BISHOP, J.M., (1983). "Cellular oncogenes and retroviruses," Ann. Rev. Biochem. 52, 301.
BLOOM, A.D., Ed. (1981). *Guidelines for Studies of Human Populations Exposed to Mutagenic and Reproductive Hazards*, (March of Dimes Birth Defects Foundation, New York).
BLOT, W.J., AKIBA, S. AND KATO, H. (1984). "Ionizing radiation and lung cancer: A review including preliminary results from a case-control study among A-bomb survivors," in *Atomic Bomb Survivor Data: Utilization and Analysis*, PRENTICE, R.L. AND THOMPSON, D.J., Eds. (SIAM, Philadelphia).
BLUMBER, P.M. (1981). "*In vitro* studies on the mode of action of the phorbol esters, potent tumor promoteps," Part 1 and 2, CRC Crit. Rev. Toxicol. 3, 152.
BLUMBERG, P.J., JAKEN, S., KONIG, B., SHARKEY, N., LEACH, K., JENG, A., AND YEH, E. (1984). "Mechanism of action of the phorbol ester tumor promoters: Specific receptors for lipophilic ligands," Biochem. Pharmacol., 33, 933.
BOICE, J.D., JR. (1979). "Multiple chest fluoroscopies and the risk of breast cancer," page 147 in *Advances in Medical Oncology, Research, and Education, Vol. 1*, MARGISON, G.P., Ed. (Pergamon Press, New York).

BOICE, J.D., JR. AND MONSON, R.R. (1977). "Breast cancer in women after repeated fluoroscopic examinations of the chest," J. Natl. Cancer Inst. 59, 832.
BOICE, J.D., JR., DAY, N.E., ANDERSEN, A., BRINTON, L.A., BROWN, P., CHOI, N.W., CLARKE, E.A., COLEMAN, M.P., CURTIS, R.E., FLANNERY, J.T., HAKAMA, M., HAKULINEN, T., HOWE, G.R., JENSEN, O.M., KLEINERMAN, R.A., MAGNIN, D., MAGNUS, K., MAKELA, K., MALKER, B., MILLER, A.B., NELSON, N., PATTERSON, C.C., PETERSSON, F., POMPE-KIRN, V., PRIMIC-ZAKELJ, M., PRIOR, P., RAVNIHAR, B., SKEET, R.G., SKJERVEN, J.E., SMITH, P.G., SOK, N., SPENGLER, R.F., STORM, H.H., TOMKINS, G.W.O., WALL, C., AND WEINSTOCK, R. (1984). "Cancer risk following radiotherapy of cervical cancer: a preliminary report," page 161 in *Radiation Carcinogenesis: Epidemiology and Biological Significance*, BOICE, J.D., JR. AND FRUAMENI, J.F., JR., Eds. (Raven Press, New York).
BOICE, J.D., JR., BEEBE, G.M. AND LAND, C.E. (1985a). "Absolute and relative time-response models in radiation risk estimation," page 22 in the *Proceedings of the Twentieth Annual Meeting of the National Council on Radiation Protection and Measurements, NCRP Proceedings, No. 6*, (National Council on Radiation Protection and Measurements, Bethesda, Maryland).
BOICE, J.D., JR., DAY, N.E., ANDERSEN, A., BRINTON, L.A., BROWN, P., CHOI, N.W., CLARKE, E.A., COLEMAN, M.P., CURTIS, R.E., AND FLANNERY, J.T. (1985b). "Second cancers following radiation treatment for cervical cancer. An international collaboration among cancer registries," J. Natl. Cancer Inst. 74, 955.
BONIVER, J., DECIEVE, A. LIEBERMAN, M., HONSIK, C., TRAVIS, M., AND KAPLAN, H.S. (1981). "Marrow-thymus interactions during radiation leukemogenesis in C57BL/Ak mice," Cancer Res. 41, 390.
BOREK, C. (1980). "X-ray induced *in vitro* neoplastic transformation of human diploid cells," Nature 283, 776.
BOREK, C. AND HALL, E.J. (1973). "Transformation of mammalian cells *in vitro* by low doses of x rays," Nature 243, 450.
BOREK, C., AND HALL, E.J. (1974). "Effect of split doses of x rays on neoplastic transformation of single cells," Nature 252, 499.
BORENFREUND, E., HIGGINS, P.J., STEINGLASS, M., AND BEINDICH, A. (1975). "Properties and malignant transformation of established rat liver parenchymal cells in culture," J. Natl. Cancer Inst. 55, 375.
BOYLAND, R. AND HARD, G.C. (1974). "Early appearance of 'transformed' cells from the kidney of rats treated with a 'single' carcinogenic dose of dimethylnitrosamine (DMN) detected by culture *in vitro*," Europ. J. Cancer 10, 177.
BORNEFF, J. (1977). "Fate of carcinogens in aquatic environments," Part 2, page 393 in *Fate of Pollutants in the Air and Water Environments*, SUFFET, I.H., Ed. (John Wiley & Sons, New York).
BOUCK, N. AND DIMAYORCA, G. (1976). "Somatic mutation as the basis for malignant transformation in BHK cells by chemical carcinogens," Nature 264, 722.

BOUTWELL, R.K. (1978). "Biochemical mechanism of tumor promotion," page 49 in *Carcinogenesis, Vol. 2., Mechanisms of Tumor Promotion and Carcinogenesis*, SLAGA, T.J., SIVAK, A. AND BOUTWELL, R.K., Eds. (Raven Press, New York).

BRADLEY, M.O., SHARKEY, N.A., KOHN, K.W., AND LAYARD, M.W. (1980). "Mutagenicity and cytotoxicity of various nitrosourease in V-79 Chinese hamster cells," Cancer Res. 40, 2719.

BROOKS, A.L., MEAD, D.K., AND PETERS, R.F. (1971). "The effect of chronic exposures to cobalt 60 on the frequency of metaphase chromosome aberrations in the liver cells of the Chinese hamster *in vivo*," Int. J. Rad. Biol. 20, 599.

BROUTY-BOYÉ, D. AND GRESSER, L. (1981). "Reversibility of the transformed and neoplastic phenotype I., Progressive reversion of the phenotype of X-ray transformed C3H/10T½ cells under prolonged treatment with interferon," Int. J. Cancer 28, 165.

BROUTY-BOYÉ, D., GRESSER, L., AND BALDWIN, C. (1979). "Reversion of the transformed phenotype to the parental phenotype by subcultivation of X-ray transformed C3H/10T½ cells at low cell density," Int. J. Cancer 24, 253.

BROWN, J.M. (1977). "The shape of the dose-response curve for radiation carcinogenesis: Extrapolation to low doses," Radiat. Res. 71, 34.

BRUCE, W.R., VARGHESE, A.J., FURRER, R., AND LAND, P.C. (1977). "A mutagen in the feces of normal humans," page 1641 in *Origins of Human Cancer*, HIATT, H.H., WATSON, J.D., AND WINSTEN, J.A., Eds. (Cold Spring Harbor Laboratory Press, Cold Spring Harbor, New York).

BURNS, F., ALBERT, R.E. AND HEIMBACK, R.D. (1968). "RBE for skin tumors and hair follicle damage in the rat following irradiation with alpha particles and electrons," Radiat. Res. 36, 225.

BURNS, F.J. AND VANDERLAAN, M. (1977). "Split-dose recovery for radiation-induced tumors in rat skin," Int. J. Radiat. Biol. 32, 135.

CALLEMAN, C.J. (1982). "A model for low dose risk estimates," page 701 in *Environmental Mutagens and Carcinogens (Proceedings of the 3rd International Conference on Environmental Mutagens)*, SUGIMURA, T., KONDO, S., AND TAKEBE, H., Eds. (University of Tokyo Press, Tokyo/Alan R. Liss. Inc., New York).

CARRANO, A.V., THOMPSON, L.H., LINDL, P.A., AND MINKLER, J.L. (1978). "Sister chromatid exchange as an indicator of mutagenesis," Nature 271, 551.

CARROLL, K.K. (1975). "Experimental evidence of dietary factors and hormone-dependent cancers," Cancer Res. 35, 3374.

CBEAP (1972). Commission on Biological Effects of Amospheric Pollutants. *Particulate Polycyclic Organic Matter.* (National Academy Press, Washington).

CHANG, E.H., FURTH, M.E., SCOLNICK, E.M., AND LOWRY, D.R. (1982). "Tumorigenic transformation of mammalian cells induced by a normal human gene homologous to the oncogene of Harvey murine sarcoma virus," Nature 297, 474.

CHEN, T.T. AND HEIDELBERGER, C. (1969). "Quantitative studies on the malignant transformation of mouse prostate cells by carcinogenic hydrocarbons *in vitro*," Int. J. Cancer 4, 166.

CLARK, C.R. AND VIGIL, C.L. (1980). "Influence of rat lung and liver homogenates on the mutagenicity of diesel exhaust particulate extracts," Toxicol. Appl. Pharmacol. 56, 110.

CLAYSON, D.G. (1984). "Modulation of the incidence of murine leukemia and lymphoma," CRC Critical Rev. Toxicol. 13, 183.

CLEAVER, J.E. (1977). "Methods for the study of excision repair of DNA damaged by physical and chemical mutagens," page 19 in *Handbook on Mutagenicity Test Procedures*, KILBEY, B.J., LEGATOR, M., NICHOLS, W., AND RAMEL, C., Eds. (Elsevier, Amsterdam).

CLIFTON, K.H. AND CROWLEY, J.J. (1978). "Effects of radiation type and dose and the rate of glucocorticoids, gonadectomy, and thyroidectomy in mammary tumor induction in mammotrophic-secreting pituitary tumor-grafted rats," Cancer Res. 38, 1507.

CLIFTON, K.H. AND SRIDHARAN, B.N. (1975). "Endocrine factors and growth," page 249 in *Cancer: A Comprehensive Treatise, Vol. 3*, BECKER, F.F., Ed. (Plenum Press, New York).

COLBURN, N.H., BRUEGGE, W.F.V., BATES, J., AND YUSPA, S.H. (1978). "Epidermal cell transformation *in vitro*," Carcinogenesis 2, 257.

COLE, J. AND ARLETT, C.F. (1978). "Methyl methanesulphonate mutagenesis in L5178Y mouse lymphoma cells," Mutat. Res. 50, 111.

COLE, L.J. AND NOWELL, P.C. (1964). "Accelerated induction of hepatomas in fast neutron-irradiated mice injected with carbon tetrachloride," Ann. New York Acad. Sci. 114, 259.

COLE, A., MEYN, R.E., CHEN, R., CORRY, P.M., AND HITTELMAN, W. (1980). "Mechanisms of cell injury," page 33 in *Radiation Biology in Cancer Research*. MEYN, R.E. AND WITHERS, H.R., Eds. (Raven Press, New York).

COMBES, R.D. AND HAVELAND-SMITH, R.B. (1982). "A review of the genotoxicity of food, drug and cosmetic colours and other azo, triphenylmethane and xanthene dyes," Mutat. Res. 98, 101.

CONKLIN, J.W., UPTON, A.C., CHRISTENBERRY, K.W. AND MCDONALD, T.P. (1963). "Comparative late somatic effects of some radiometric agents and x rays," Radiat. Res. 19, 156.

COUCH, D.B., FORBES, N.L., AND HSIE, A.W. (1978). "Comparative mutagenicity of alkylsulfate and alkylsulfonate derivatives in Chinese hamster ovary cells," Mutat. Res., 57, 217.

COURT-BROWN, W.M. AND DOLL, R. (1957). "Leukaemia and aplastic anaemia in patients irradiated for ankylosing spondylitis," Medical Research Council, Special Series No. 295 (H.M.S.O., London).

CRONKITE, E.P., BULLIS, J., INOUE, T., AND DREW, R.T. (1984). "Benzene inhalation produces leukemia in mice," Toxicol. Appl. Pharmacol. 75, 358.

CROUCH, E.A.C. (1983a). "Uncertainties in interspecies extrapolations of carcinogenicity," Env. Hlth. Perspect. 50, 321.

CROUCH, E.A.C. (1983b). "Carcinogenic risk assessment — the consequences of believing models," Basic Life Sci. 24, 653.

CROUCH, E.A.C., AND WILSON, R. (1979). "Interspecies comparison of carcinogenicity," J. Tox. Environ. Hlth. 5, 1095.
CROUCH, E.A.C., AND WILSON, R. (1981). "Regulation of carcinogens," Risk Analysis 1, 47.
CRUMP, K.S. (1981). "An improved procedure for low-dose carcinogenic risk assessment from animal data," J. Environ. Pathol. Toxicol. 5, 675.
CRUMP, K.S., HOEL, D., LANGLEY, C., AND PETO, R. (1976). "Fundamental carcinogenic processes and their implications for low dose risk assessment," Cancer Res., 36: 2973.
CUTLER, S.J. AND YOUNG, J.L., JR., Eds. (1975). *Third National Cancer Survey: Incidence Data*, DHEW (NIH) Publ. (Government Printing Office, Washington), Natl. Cancer Inst. Monogr. 41, 75–787.
DAGLE, G.E. AND SANDERS, C.L. (1982). "Radionuclide injury to the lung," in *Pulmonary Toxicology*, HOOK, G., Ed. (Raven Press, New York).
DEDRICK, R.L. (1973). "Animal scale-up," J. Pharmacokinetics and Biopharmaceutics 1, 435.
DEGRAEVE, N. (1981). "Carcinogenic, teratogenic and mutagenic effects of cadmium," Mutat. Res. 86, 115.
DEILA PORTA, G., RAPPAPORT, H. AND SAFFIOTTI, U. (1956). "Induction of melanotic lesions during skin carcinogenesis in hamsters," AMA Archives of Pathology 61, 305.
DEMARS, R. AND JACKSON, J.L. (1977). "Mutagenicity detection with human cells," J. Environ. Pathol. Toxicol. 1, 55.
DIETZ, E.K., RAMSEY, J.C., AND WATANABE, P.G. (1983). "Experimental studies to human risk," Environ. Health Perspect. 52, 9.
DIPAOLO, J.A., DONOVAN, P.J., AND NELSON, R.L. (1971a). "Transformation of hamster cells *in vitro* by polycyclic hydrocarbons without cytotoxicity," Proc. Natl. Acad. Sci. 68, 2958.
DIPAOLO, J.A., DONOVAN, P.J. AND NELSON, R.L. (1971b). "*In vitro* transformation of hamster cells by polycyclic hydrocarbons: factors influencing the number of cells transformed," Nature New Biology 230, 240.
DIPAOLO, J.A., TAKANO, K., AND POPESCU, N.C. (1972). "Quantitation of chemically induced neoplastic transformation of Balb/c 3T3 cloned cell lines," Cancer Res. 32, 2686.
DOBIAS, L. (1980). "Human blood mutagenicity for *Salmonella typhimurium* tester strains after oral application of entizol," Mutat. Res. 77, 357.
DOLL, R. (1970). "Cancer and aging: The epidemiologic evidence," Oncology 5, 1.
DOLL, R. (1978). "An epidemiological perspective of the biology of cancer," Cancer Res. 38, 3573.
DOLL, R. AND PETO, R. (1981). "The cause of cancer: quantitative estimates of avoidable risk of cancer in the United States today," J. Natl. Cancer Inst. 66, 1192.
EBBESEN, P. (1977). "Effect of age of non-skin tissues on susceptibility of skin grafts to 7,12-dimethylbenz[a]anthracene (DMBA) carcinogenesis in Balb/c mice and effect of skin graft on susceptibility of surrounding recipient skin to DMBA," J. Natl. Cancer Inst. 58, 1057.

EHLING, U.H., AVERBECK, D., CERUTTI, P.A., FRIEDMAN, J., GRIEM, H., KOLBYE, A.C. JR., AND MENDELSOHN, M.L. (1983). "Review of the evidence for the presence or absence of thresholds in the induction of genetic effects by genotoxic chemicals," International Commission for Protection Against Environmental Mutagens and Carcinogens, ICPEMC Publication No. 10., Mutat. Res. **123**, 281.

EHRENBERG, L. (1979). "Risk assessment of ethylene oxide and other compounds in Banbury Report 1," page 157 in *Assessing Chemical Mutagens: The Risk to Humans,* MCELHENY, V.K. AND ABRAMSON, S., Eds. (Cold Spring Harbor Laboratory Press, Cold Spring Harbor, New York).

ELKIND, M.M. AND HAN, A. (1979). "Neoplastic transformation and dose fractionation; does repair of damage play a role," Radiat. Res. **79**, 233.

EPA (1984). Environmental Protection Agency. *Carcinogen Assessment of Coke Oven Emissions,* United States Environmental Protection Agency, Office of Health and Environmental Assessment, EPA-600/6-82-003F, Final Report, (Environmental Protection Agency, Washington).

EPA (1986a). Environmental Protection Agency. *Guidelines for mutagenicity risk assessment,* Federal Register 51(185):33991 (Government Printing Office, Washington).

EPA (1986b). Environmental Protection Agency. *Guidelines for exposure assessment,* Federal Register 51(185):34041 (Government Printing Office, Washington).

EVANS, H.J. AND SCOTT, D. (1969). "The induction of chromosome aberrations by nitrogen mustard and its dependence on DNA synthesis," Proc. R. Soc. Brit. **173**, 491.

FARBER, E. (1984). "Cellular biochemistry of the stepwise development of cancer with chemicals: G.H.A. Clowes Memorial Lecture," Cancer Res. **44**, 5463.

FELDMAN, G., REMSEN, J., WANG, T.V., AND CERUTTI, P. (1980). "Formation and excision of covalent deoxyribonucleic acid adducts of benzo[a]pyrene, 4,5-epoxide and benzo[a]pyrene diol epoxide I in human lung cells A549," Biochem. **19**, 1095.

FERNANDEZ, A., MONDAL, S., AND HEIDELBERGER, C. (1980). "Probabilistic view of the transformation of cultured C3H/10T1/2 mouse embryo fibroblasts by 3-methylchoanthrene," Proc. Natl. Acad. Sci, **77**, 7272.

FIALKOW, P.J. (1977). "Clonal origin of human tumors," Biochem. Biophys. Acta **458**, 283.

FRANK, J.P. AND WILLIAMS, J.R. (1982). "X-ray induction of persistent hypersensitivity to mutation," Science **216**, 307.

FREEDMAN, V.H. AND SHIN, S. (1977). "Isolation of human diploid cell variants with enhanced colony-forming efficiency in semisolid medium after a singlestep chemical mutagenesis," J. Natl. Cancer Inst. **58**, 1873.

FRIEDMAN, J. AND HUBERMAN, E. (1980). "Postreplication repair and the susceptibility of Chinese hamster cells to cytotoxic and mutagenic effects of alkylating agents," Proc. Natl. Acad. Sci. **77**, 6072.

FURTH, J. (1975). "Hormones as etiological agents in neoplasia," page 75 in *Cancer* Vol. 1. BECKER, F., Ed. (Plenum Press, New York).

FUSENIG, N.E., AMER, S.M., BONKAMP, P., AND WORST, P.K.M. (1978). "Characteristics of chemically transformed mouse epidermal cells *in vitro* and *in vivo*," Bull. Cancer **65**, 271.

GARDNER, W.U. AND DOUGHERTY, T.F. (1944). "The leukemogenic action of estrogens in hybrid mice," Yale, J. Biol. Med. **17**, 75.

GART, J.J., DIPAOLO, J.A., AND DONOVAN, P.J. (1979). "Mathematical models and the statistical analyses of cell transformation experiments," Cancer Res. **39**, 6069.

GAYLOR, D.W. AND KODELL, R.L. (1980). "Linear interpolation algorithm for low dose risk assessment of toxic substances," J. Environ. Pathol. Toxicol. **4**, 305.

GOERTTLER, K. AND LAEHRKE, H. (1976). "Diaplacental carcinogenesis: initiation with the carcinogens dimethyl-benzanthracene (DMBA) and urethane during fetal life and postnatal promotion with the phorbol ester TPA in a modified 2-stage Berenblum/Mottram experiment," Virch. Arch. Path. Anat. Histol. **372**, 29.

GOLD, L.S., SAWYER, C.B., MAGAW, R., BLACKMAN, G.M., DE VECIANA, M., LEVINSON, R., HOOPER, N.K., HAVENDER, W.R., BERNSTEIN, L., PETO, R., *et al.*,(1984). "A carcinogenic potency database of the standardized results of animal bioassays," Env. Health Perspect **58**, 9.

GOLDSTEIN, A., ARONOW, L., AND KALMAN, S.M., Eds. (1974). *Principles of drug action: The basis of pharmacology*, 3rd ed. (Harper and Row, Inc., New York).

GOMEZ-ARROYO, S., ALTAMIRANO, M., AND VILLALOBOS-PIETRINI, R. (1981). "Sister-chromatid exchanges induced by some chromium compounds in human lymphocytes *in vitro*," Mutat. Res. **90**, 425.

GONZALEZ, F.J., JAISWAL, A.K., AND NEBERT, D.W. (1986). "P 450 Genes: Evolution, regulations and relationship to human cancer," in *Cold Spring Harbor Symposium on Quantitative Biology 51* (Cold Spring Harbor Laboratory Press, Cold Spring Harbor, New York).

GOTH, R. AND RAJEWSKY, M.F. (1974). "Persistence of 6-ethylguanine in rat system specific carcinogenesis by ethylnitrosourea," Proc. Natl. Acad. Sci. **71**, 639.

GROOPMAN, J.D., HAUGEN, A., GOODRICH, G.R., WOGAN, G.N., AND HARRIS, C.C. (1982). "Quantitation of aflatoxin B_1-modified DNA using monoclonal antibodies, Cancer Res. **42**, 3120.

GROSOVSKY, A.J. AND LITTLE, J.B. (1985). "Evidence for linear response for the induction of mutations in human cells by x-ray exposure below 10 rads," Proc. Natl. Acad. Sci. **82**, 092.

GOTH-GOLDSTEIN, R. AND BURKI, H.J. (1980). "Ethylnitrosourea-induced mutagenesis in asynchronous and synchronous Chinese hamster ovary cells," Mutat. Res. **69**, 127.

GUERRERO, I., CALZADA, P., MAYER, A., AND PELLICER, A. (1984). "A molecular approach to leukemogenesis: mouse lymphomas contain an activated c-ras oncogene," Proc. Natl. Acad. Sci. **81**, 202.

GUILLINO, P.M., PETTIGREW, H.M., AND GRANTHAM, F.H. (1975). "N-Nitrosomethylurea as mammary gland carcinogen in rats," J. Natl. Cancer Inst. **54**, 401.

HABER, D.A. AND THILLY, W.G. (1978). "Morphological transformation of C3H 10T1/2 cells subcultured at low cell densities," Life Sciences 2, 1663.
HABER, D.A., FOX, D.A., DYNAN, W.S., AND THILLY, W.G. (1977). "Cell density dependence of focus formation in the C3H 10T1/2 transformation assay," Cancer Res. 37, 1644.
HAN, A. AND ELKIND, M.M. (1979). "Transformation of mouse C3H 10T1/2 cells by single fractionated doses of x rays and fission-spectrum neutrons," Cancer Res. 39, 123.
HARAN-GHERE, N. (1978). "Target cells involved in radiation and radiation leukemia virus leukemogenesis," page 79 in *Radiation-Induced Leukemogenesis and Related Viruses*, INSERM Symposium No. 4, DUPLAN, J.F., Ed. (Elsevier Press, New York)
HARRIS, C.C., VAHAKANGAS, K., NEWMAN, M.J., TRIVERS, G.E., SHAMSUDDIN, A., SINOPOLI, N., MANN, D.L., AND WRIGHT, W.E. (1985). "Detection of benzo[a]pyrene diol epoxide-DNA adducts in peripheral blood lymphocytes and antibodies to the adducts in serum from coke oven workers," Proc. Natl. Acad. Sci. 82, 6672.
HASEMAN, J.K. (1985). "Issues in carcinogenicity testing: dose selection," Fund. Appl. Toxicol. 5, 66.
HEIDELBERGER, C., FREEMAN, A.E., PIENTA, R.K., SIVAK, A., BERTRAM, J.S., CASTO, B.C., DUNKEL, V.C., FRANCIS, M.W., KAKUNAGA, T., LITTLE, J.B., AND SCHECHTMAN, L.M. (1983). "Cell transformation by chemical agents: a review and analysis of the literature," A report of the EPA Genetox Program, Mutat. Res. 114, 283.
HENNINGS, H. MICHALE, D., AND PATERSON, E. (1973). "Enhancement of skin tumorigenesis by single application of croton oil before or soon after initiation by urethane," Cancer Res. 33, 3130.
HERBERMAN, R.B. (1984) "Possible role of natural killer cells and other effector cells in immune surveillance against cancer," J. Invest. Dermatol, 83, 1375.
HICKS, R.M., WAKEFIELD, S.J., AND CHOWANIEC, J. (1975). "Evaluation of a new model to detect bladder carcinogens or co-carcinogens: results obtained with saccharine, cyclamate and cyclophosphamide," Chem.-Biol. Interactions 11, 224.
HICKS, R.M., CHOWANIEC, J., AND WAKEFIELD, S.J. (1978). "Experimental induction of bladder tumors by a two-stage system," in *Mechanisms of Tumor Promotion and Cocarcinogenesis*, SLAGA, T.J., SIVAK, A. AND BOUTWELL, R.K., Eds., (Raven Press, New York).
HILDEBRAND, C.E., GONZALES, F.J., MCBRIDE, O.W., AND NEBERT, D.W. (1985). "Assignment of the human 2,3,5,8-tetrachlorodibenzo-p-dioxin-inducible cyto chrome P_1-450 gene to chromosome 15," Nucl. Acids Res. 13, 2009.
HILL, C.K., HAN, A., AND ELKIND, M.M. (1984). "Fission-spectrum neutrons at a low dose rate enhance neoplastic transformation in the linear, low dose region (0–10 cGy)," Int. J. Radiat. Biol. 46: 11.
HIRASHIMA, K., BESSHO, M., HAYATA, I., NARA, N., KAWASE, Y., AND OHTANI, M. (1982). "Transformation of bone marrow stem cells and radiation induced myeloid leukemia in mice," page 85 in *Proceedings of the workshop on*

Tritium Radiobiology and Health Physics (National Institute of Radiological Sciences, Chiba, Japan).

HOEL, D.G., KAPLAN, N.L., AND ANDERSON, M.W. (1983). "Implication of nonlinear kinetics on risk estimation in carcinogenesis," Science **219**, 1023.

HOLLSTEIN, M., MCCANN, J., ANGELOSANTO, F.A., AND NICHOLS, W.W. (1979). "Short term tests for carcinogens and mutagens," Mutat. Res. **65**, 289.

HOLM, L.E., EKLUND, G., AND LUNDELL, G. (1980a). "Incidence of malignant thyroid tumors in humans after exposure to diagnostic doses of iodine-131: II. Estimation of thyroid gland size, thyroid radiation dose, and predicted versus observed number of malignant thyroid tumors," J. Natl. Cancer Inst. **65**, 1221.

HOLM, L.E, DAAHLQUIST, I., ISRAELSSON, A., AND LUNDELL, G. (1980b). "Malignant thyroid tumors after iodine-131 therapy," N. Engl. Med. J. **303**, 188.

HOWE, G.R. (1984). "Epidemiology of radiogenic breast cancer," pp. 119 in *Radiation Carcinogenesis: Epidemiology and Biological Significance*, BOICE, J.D., JR. AND FRAUMENI, J.F., JR., Eds. (Raven Press, New York).

HSU, T.C., POIRIER, M.C., YUSPA, S.H., GRUNBERGER, D., WEINSTEIN, I.B., YOLKEN, R.H., AND HARRIS, C.C. (1981). "Measurement of benzo[a]pyrene-DNA adducts by enzyme immunoassays and radioimmunoassay," Cancer Res. **41**, 1091.

HUANG, S.L. AND LIEBERMAN, M.W. (1978). "Induction of 6-thioguanine resistance in human cells treated with N-acetoxy-2-acetylaminofluorene," Mutat. Res. **57**, 349.

HUBERMAN, E. AND SACHS, L. (1966). "Cell susceptibility to transformation and cytotoxicity by the carcinogenic hydrocarbon benzo[a]pyrene," Proc. Natl. Acad. Sci. **56**, 1123,

HUBERMAN, E., MAGER, R., AND SACHS, L. (1976). "Mutagenesis and transformation of normal cells by chemical carcinogens," Nature **204**, 360.

HUGGINS, C. AND SUGIYAMA, T. (1966). "Induction of leukemia in rat by pulse-doses of 7, 12-dimethylbenz[a]anthracene," Proc. Natl. Acad. Sci. **55**, 74.

HUGHES, T.J., PELLIZZARI, E., LITTLE, L., SPARACINO, C., AND KOLBER, A. (1980). "Ambient air pollutants: Collection, chemical characterization and mutagenicity testing," Mutat. Res. **76**, 51.

HUTCHISON, G.B. (1968). "Leukemia in patients with cancer of the cervis uteri treated with radiation. A report covering the first 5 years of an international study," J. Natl. Cancer Inst. **40**, 951.

IANNACONE, P.M., GARDNER, R.L. AND HARRIS, H. (1978). "The cellular origin of chemically induced tumors," J. Cell Sci. **29**, 249.

IARC. (1980). International Agency for Research on Cancer. *Longer-term and Short-term Screening Assays for Carcinogen in a Critical Appraisal*, Supplement 2. (International Agency for Research on Cancer, Lyon).

IARC. (1982). International Agency for Research on Cancer. *On the Evaluation of the Carcinogenic Risk of Chemicals to Humans, Chemicals, Industrial Processes, and Industries Associated with Cancer in Humans*, Supplement 4. (International Agency for Research on Cancer, Lyon).

IARC/NCI/EPA (1985). International Agency for Research on Cancer/National Cancer Institute/Environmental Protection Agency. "Cellular and molecular mechanisms of cell transformation and standardization of transformation assays of established cell lines for the prediction of carcinogenic chemicals: overview and recommended protocols," Cancer Res. 45, 2395.

IBALL, J. (1939). "The relative potency of carcinogenic compounds," Am. J. Cancer 35, 188.

ICHIMARU, M., ISHIMARU, T., AND BELSKY, J.L. (1978). "Incidence of leukemia in atomic bomb survivors belonging to a fixed cohort in Hiroshima and Nagasaki, 1950-71. Radiation dose, years after exposure, age at exposure, and type of leukemia," Japan J. Radiat. Res. 19, 262.

ICRP (1974). International Commission on Radiological Protection. *Report of Task Group on Reference Man*, ICRP Publication 23 (Pergamon Press, New York).

ICRP (1977). International Commission on Radiological Protection. *Recommendations of the International Commission on Radiological Protection*, ICRP Publication 26, Annals of the ICRP, Vol. 1, No. 3 (Pergamon Press, New York).

ICRP (1985). International Commission on Radiological Protection. "Statement from the 1985 Paris Meeting of the International Commission on Radiological Protection," Health Phys. 48, 828.

ICRU (1970). International Commission on Radiation Units and Measurements. *Linear Energy Tranfer*, ICRU Report No. 16 (International Commission on Radiation Units and Measurements, Bethesda, Maryland).

ICRU (1980). International Commission on Radiation Units and Measurements. *Radiation Quantities and Units*, ICRU Report No. 33 (International Commission on Radiation Units and Measurements, Bethesda, Maryland).

IMAMURA, N., NAKANO, M., KAWASE, A., KAWAMURA, Y., AND YOKORO, K. (1973). "Synergistic action of N-nitrosobutylurea and azathioprine in induction in C57BL mice," Gann 64, 493.

INUI, N., KAKETOMI, M., AND NISHI, Y. (1979). "Mutagenic effects of AF-2, a food additive, on embryonic cells of the Syrian golden hamster on transplacental application," Mutat. Res. 41, 351.

IRINO, S., SEZAKI, T., MORIYA, J., AND HIRAKI, K. (1968). "Organ distribution of 3-methylcholanthrene after repeated painting and oral administration in relation to the induction of leukemia in mice," Gann 59, 223.

IRLG (1979). Interagency Regulatory Liaison Group. "Work Group on Risk Assessment, Scientific bases for identification of potential carcinogens and estimation of risks," J. Natl. Cancer Inst. 63, 241.

IRR, J.D. AND SNEE, R.D. (1979). "Statistical evaluation of mutagenicity in the CHO/HGPRT system," page 263 in *Banbury Report 2, Mammalian Cell Mutagenesis: The Maturation of Test Systems*. HSIE, A.W., O'NEIL, J.P., AND MCELHENY, V.K., Eds. (Cold Spring Harbor Laboratory Press, Cold Spring Harbor, New York).

ISHIMARU, T., HOSHINO, T., ICHIMARU, M., OKADA, H., TOMIYASU, T., TSUCHIMOTO, T., AND YAMAMOTO, T. (1969). *Leukemia in Atomic Bomb Survivors, Hiroshima-Nagasaki, October 1, 1950 - September 30, 1966*, ABCC TR 25-

69. (Radiation Effects Research Foundation, Hiroshima).
ISHIMARU, T., CIHAK, R.W., LAND, C.E., STEER, A., AND YAMADA, A. (1975). "Lung cancer at autopsy in A-bomb survivors and controls, Hiroshima and Nagasaki, 1961–1970: II. Smoking, occupation, and A-bomb exposure," Cancer 36, 1723.
JACOBS, L. AND DEMARS, R. (1978). "Quantification of chemical mutagenesis in diploid human fibroblasts: induction of azaguanine-resistant mutants by N-methyl N'-nitro-N-nitrosoguanidine," Mutat. Res. 53, 29.
JAISWAL, A.K., GONZALEZ, F.J., AND NEBERT, D.W. (1985a). "Human P_1-450 gene sequence and correlation of mRNA with genetic differences in benzo[a]pyrene metabolism," Nucl. Acids Res. 13, 4503.
JAISWAL, A.K., NEBERT, D.W., AND EISEU, H.J. (1985b). "Comparison of aryl hydrocarbon hydroxase and acetanilide 4-hydroxylase induction of polycyclic aromatic compounds in human and mouse cell lines," Biochem. Pharmacol. 34, 2721.
JAKOBY, W.B., BEND, J.R., AND CALDWELL, J. (1982). *Metabolic basis of detoxication: Metabolism of functional groups* (Academic Press, New York).
JENSEN, D. (1981). "Low-dose mutagenicity in mammalian cells," thesis, (University of Stockholm, Stockholm, Sweden).
JENSEN, D., AND RAMEL, C. (1980). "Relationship between chemical damage of DNA and mutations in mammalian cells. I. Dose-response curves for the induction of 6-thioguanine resistant mutants by low doses of monofunctional alkylating agents, x rays and UV radiation in V79 Chinese hamster cells, Mutat. Res. 73, 339.
JOSTES, R.F., JR. (1981). "Sister-chromatid exchanges but not mutations decrease with time in arrested Chinese hamster ovary cells after treatment with ethylnitrosourea," Mutat. Res. 91:371.
JULL, J.W. (1976). "Endocrine aspects of carcinogenesis," page 52 in *Chemical Carcinogens*, SEARLE, C.E., Ed. (American Chemical Society, Washington).
KAKUNAGA, T. (1978). "Neoplastic transformation of human diploid fibroblast cells by chemical carcinogens," Proc. Natl. Acad. Sci. 75, 1334.
KAPLAN, H.S. (1967). "On the natural history of the murine leukemias: presidential address," Cancer Res. 27, 1325.
KAPLAN, H.S. (1984). "Retroviral aetiology of 'virus-free' animal leukemias and lymphomas," page 1 in *Mechanisms of Viral Leukaemogenesis*, GOLDMAN, J.M. AND JARETT, O. Eds. (Churchill Livingston, New York).
KAPLAN, H.S. AND BROWN, M.B. (1952). "A quantitative dose response study of lymphoid-tumor development in irradiated C57 black mice," J. Natl. Cancer Inst. 13, 185.
KATO, H. AND SCHULL, W.J. (1982). "Studies of the mortality of A-bomb survivors. 7. Mortality, 1950–1978. Part 1. Cancer mortality," Radiat. Res. 90, 395.
KATSUTA, H. AND TAKAOKA, T. (1975). "Chemical carcinogenesis of mammalian epithelial cells in tissue culture," Cancer Res. 17, 59.
KAUFFMAN, S.L., ALEXANDER, L., AND SAAS, L. (1979). "Histological and ultrastructural features of the Clara cell adenoma of the mouse lung," Lab. Invest. 40, 708.

KENNEDY, A.R. (1984). "Promotion and other interactions between agents in the induction of transformation *in vitro* in fibroblast-like cell culture systems," in *Mechanisms of Tumor Promotion, Vol. 3. Tumor Promotion and Carcinogenesis In Vitro*, SLAGA, T.J., Ed. (CRC Press, Inc., Cleveland, Ohio).

KENNEDY, A.R., AND LITTLE, J.B. (1980). "Investigations of the mechanisms for enhancement of radiation transformation *in vitro* by 12-0-tetradecanoylphorobo-13-acetate," Carcinogenesis 1, 1039.

KENNEDY, A.R., FOX, M., MURPHY, G., AND LITTLE, J.B. (1980). "On the relationship between x-ray exposure and malignant transformation in C3H 10T1/2 cells," Proc. Natl. Acad. Sci. 77, 7262.

KENNEDY, A.R., CAIRNS, J., AND LITTLE, J.B. (1984). "Timing of the steps in transformation of C3H 10T1/2 cells by x-irradiation," Nature 307, 85.

KIRSCHBAUM, A. (1951). "Rodent leukemia: recent biological studies," Cancer Res. 11, 741.

KNEIP, T.J., DAISEY, J.M., SOLOMON, J.J., AND HERSHMAN, R.J. (1983). "N-nitroso compounds: Evidence for theirs presence in airborne particles," Science 21, 1045.

KOURI, R.E. AND NEBERT, D.W. (1977). "Genetic regulation of susceptibility to polycyclic-hydrocarbon-induced tumors in the mouse," page 811 in *Origins of Human Cancer*. HIATT, H.H., WATSON, J.D., AND WINSTEN, J.A., Eds. (Cold Spring Harbor Laboratory Press, Cold Spring Harbor, New York).

KOURI, R.E., MCKINNEY, C.E., SLOMIANY, D.J., SNODGRASS, D.R., WRAY, N.P., AND MCLEMORE, T.L. (1982). "Positive correlation between high aryl hydrocarbon hydroxylase activity and primary lung cancer as analyzed in cryopreserved lymphocytes," Cancer Res. 42, 5030.

KREWSKI, D. AND BROWN, C. (1981). "Carcinogenic risk assessment: a guide to the literature," Biometrics 37, 353.

KREWSKI, D. AND VAN RYZIN, J. (1981). "Dose response models for quantal response toxicity data," page 201 in *Statistics and Related Topics*, SXORGO, J., DAWSON, D., RAO, J.N.K., AND SALEH, E., Eds. (North Holland, New York).

KROES, R., SONTAG, J.M., SELL, S., WILLIAMS, G.M., AND WEISBURGER, J.H. (1975). "Elevated concentrations of serum alpha-fetoprotein in rats with chemically induced liver tumors," Cancer Res. 35, 1214.

KUCEROVA, M., POLIVKOVA, Z., AND BATORA, J. (1979). "Comparative evaluation of the frequency of chromosomal aberrations and the SCE numbers in peripheral lymphocytes of workers occupationally exposed to vinyl chloride monomer," Mutat. Res. 67, 97.

KULESZ-MARTIN, M., KOEHLER, B., HENNINGS, H., AND YUSPA, S.H. (1981). "Quantitative assay for carcinogen altered mouse epidermal cells," Cancer Res. 41, 79.

KURELEC, B., MATIJASEVIC, Z., RIJAVEC, M., ALACEVIC, M., BRITVID, S., MULLER, W.E.G., AND ZAHN, R.K. (1979). "Induction of benzo[a]pyrene monooxygenase in fish and the Salmonella test as a tool for detecting mutagenic/carcinogenic xenobiotics in the aquatic environment," Bull. Environ. Contam. Toxicol. 21, 799.

LAERUM, O.D. AND RAJEWSKY, M.F. (1975). "Neoplastic transformation of fetal rat brain cells in culture after exposure to ethylnitrosurea *in vivo*," J. Natl. Cancer Inst. 55, 1177.

LAND, C.E., AND TOKUNAGA, M. (1984). "Induction period," page 421 in *Radiation Carcinogenesis: Epidemiology and Biological Significance*, BOICE, J.D., JR. AND FRAUMENI, J.F., JR., Eds. (Raven Press, New York).

LAND, C.E., BOICE, J.D., JR., SHORE, R.E., NORMAN, J.E., AND TOKUNAGA, M. (1980). "Breast cancer risk from low-dose exposures to ionizing radiation: Results of parallel analysis of three exposed populations of women," J. Natl. Cancer Inst. 65, 353.

LAND, H., PARADA, L.F., AND WEINBERG, R.A. (1983a). "Cellular oncogenes and multistep carcinogenesis," Science 222, 771.

LAND, H., PARADA, L.G., AND WEINBERG, R.A. (1983b). "Tumorigenic conversion of primary embryo fibroblasts requires at least two cooperating oncogenes," Nature 304, 596.

LATT, S.A., ALLEN, J., BLOOM, S.E., CARRANO, A., FALKE, E., KRAM, D., SCHNEIDER, E., SCHRECK, R., TICE, R., WHITFIELD, B. AND WOLFF, S. (1981). "Sister-chromatid exchanges: A report of the Gene-Tox program," Mutat. Res. 87, 17.

LEE, W. AND YUMMANS, H.D., JR. (1970). Doses in the brain, eyes, and thyroid of children from 60–110 Kvp x-ray epilation for tinea capitis," Phys. Med. Biol. 15, 219.

LEGATOR, M.S., CONNOR, T.H., AND STOCKEL, M. (1975). "Detection of mutagenic substances in the urine and blood of man," Annals N.Y. Acad. Sci. 269, 16.

LEHMAN, A.R. AND BRIDGES, B.A. (1977). "DNA Repair," Essays Biochem. 13, 71.

LEWIS, E.B. (1957). "Leukemia and ionizing radiation," Science 112, 965.

LINDAHL, T. (1982). "DNA repair enzymes," Ann. Rev. Biochem. 51, 61.

LITTLE, J.B., MCGRANDY, R.B., AND KENNEDY, A.R. (1978). "Interactions between Polonium 210 irradiation, benzo[a]pyrene, and 0.9% NaCl solution instillations in the induction of experimental lung cancer," Cancer Res. 38, 1929.

LITTLEFIELD, N.A., FARMER, J.H., AND GAYLOR, D.W. (1979). "Effects of dose and time in a long-term, low-dose carcinogenic study," J. Environ. Pathol. Toxicol. 3, 17.

LLOYD, D.C. AND PURROTT, R.J. (1981). "Chromosome aberration analysis in radiological protection dosimetry," Radiat. Prot. Dos. 1, 19.

LOEB, L.A. AND KUNKEL, T.A. (1982). "Fidelity of DNA synthesis," Ann. Rev. Biochem. 52, 429.

LOEB, L., ERNSTER, V.L. WARNER, K.E., ABBOTTS, J., AND LASZEL, J. (1984). "Smoking and lung cancer; an overview," Cancer Res. 44, 5940.

LOEWE, W.E. AND MENDELSOHN, E. (1981). "Revised dose estimates at Hiroshima and Nagasaki," Health Phys. 41, 663.

LOPER, J.C. (1980). "Mutagenic effects of organic compounds in drinking water," Mutat. Res. 76, 241.

LUNDE, G. AND BJORSETH, A. (1977). "Polycyclic aromatic hydrocarbons in long-range transported aerosols," Nature 268, 518.

LUNDIN, F.E., JR., WAGONER, J.K., AND ARCHER, V.E. (1971). *Radon Daughter Exposure and Respiratory Cancer: Quantitative and Temporal Aspects*, NIOSH-NIEHS Joint Monogr. No. 1 (National Technical Information Service, Springfield, Virginia).

LUTZER, M.A. (1977). "Nosology among the neoplastic geneodermatoses," page 145 in *Genetics of Human Cancer*, MULVIHILL, J.J., MILLER, R.W., AND FRAUMENI, J.F., JR., Eds. (Raven Press, New York).

MACMAHON, B. (1962). "Pre-natal x-ray exposure and childhood cancer," J. Natl. Cancer Inst. 28, 1173.

MACMAHON, B. (1981). "Childhood cancer and prenatal irradiation," page 223 in *Cancer Achievements, Challenges, and Prospects for the 1980s, Vol. 1*, BURCHENAL, J.H. AND OETTGEN, H.F., Eds. (Grune & Stratton, New York).

MAGEE, P.N., MONTESANO, R., AND PREUSSMANN, R. (1976). "N-nitroso compounds and related carcinogens," page 491 in *Chemical Carcinogens*. SEARLE, C.E., Ed. (American Chemical Society, Washington).

MAHER, V., ROWAN, L.A., SILINSKAS, K., KATELEY, S., AND MCCORMICK, J. (1982). "Frequency of UV-induced neoplastic transformation of diploid human fibroblasts is higher in xeroderma pigmentosum cells than in normal cells," Proc. Natl. Acad. Sci. 79, 2613.

MALDAGUE, P. (1969). "Comparative study of experimentally induced cancer of the kidney in mice and rats with x-rays," page 439 in *Radiation-Induced Cancer*, IAEA STI/PUB/228 (International Atomic Energy Agency, Vienna).

MALTONI, C. (1977). "Vinyl chloride carcinogenicity: an experimental model for carcinogenesis studies," page 119 in *Origins of Human Cancer*, HIATT, H.H., WATSON, J.D. AND WINSTEN, J.A., Eds. (Cold Spring Harbor Laboratory Press, Cold Spring Harbor, New York).

MALTONI, C., CONTI, B., AND COTTI, G. (1983). "Benzene: A multipotential carcinogen. Results of long-term bioassays performed at the Bologna Institute of Oncology," Am. J. Ind. Med. 4, 589.

MANTEL, N., BOHIDAR, N.R., BROWN, C.C., CIMINERA, J.L., AND TUKEY, J.W. (1975). "An improved Mantel-Bryan procedure for 'safety' testing of carcinogens." Cancer Res. 35, 865.

MARCH, H.C. (1944). "Leukemia in radiologists," Radiology 4, 275.

MARCHOK, A.C., RHOTON, J.L., GRIESEMER, R.A., AND NETTESHEIM, P. (1977). "Increased *in vitro* growth capacity of tracheal epithelium exposed *in vivo* to 7,12-dimethylbenz[a]anthracene," Cancer Res. 37, 1811.

MARCHOK, A.C., RHOTON, J.C., AND NETTESHEIM, P. (1978). "*In vitro* development of oncogenicity in cell lines established from tracheal epithelium pre-exposed *in vivo* to 7,12 dimethylbenz[a]anthracene," Cancer Res. 38, 2030.

MARSHALL, J.M. AND GROER, P.G. (1977). "A theory of the induction of bone cancer by alpha radiation," Radiat. Res. 71, 149.

MAYNEORD, W.V. (1978). "The time factor in carcinogenesis," Health Phys. 34, 297.

MAYS, C.W., ROWLAND, R.E., AND STEHNEY, A.F. (1985). "Cancer risk from the lifetime intake of radium and uranium isotopes," Health Phys. 48, 635.

MCANN, J., CHOI, E., YAMASAKI, E., AND AMES, B.N. (1975). "Detection of carcinogens as mutagens in the *Salmonella*/microsome test: Assay of 300 chemicals," Proc. Natl. Acad. Sci. 72, 5135.

McCormick, J.J., Silinskas, K.L., and Maher, V.M. (1980). "Transformation of diploid human fibroblasts by chemical carcinogens," page 491 in *Carcinogenesis, Fundamental Mechanisms and Environmental Effects*, Pullman, B. Ts'o, P.O.P. and Gelboin, H., Eds. (D. Reidel Publishing Company, Boston).

McGregory, J.F. (1979). "Association of malignancy with rapid growth in early lesions induced by irradiation of rat skin," J. Natl. Cancer Inst. 62, 1043.

Medina, D. (1976). "Preneoplastic lesions in murine mammary cancer," Cancer Res. 36, 2589.

Medina, D. (1977). "Tumor formation in preneoplastic mammary nodule lines in mice treated with nafoxidine, testosterone, and 2-bromo-α-ergocryptine," J. Natl. Cancer Inst. 58, 1107.

Melief, C.J.M. and Schwartz, R.S. (1975). "Immunocompetence and malignancy," page 121 in *Cancer: A Comprehensive Treatise I.*, Becker, F.F., Ed. (Plenum Press, New York).

Mendelsohn, M.L., Bigbee, W.L., Branscomb, E.W., and Stanatoganno- poulos, G. (1980). "The detection and sorting of rare sickle-hemoglobin containing cells in normal human blood," Cytometry 4, 311.

Metalli, P., Covelli, V., DiPaola, M., and Silini, G. (1974). "Dose incidence data for mouse reticulum cell sarcoma," Radiat. Res. 59, 21.

Miller, R. and Hall, E.J. (1978). "X-ray dose fractionation and oncogenic formations in cultured mouse embryo cells," Nature 272, 58.

Miller, R.C., Hall, E.J., and Rossi, H.H. (1979). "Oncogenic transformation in mammalian cells *in vitro* with split doses of x rays," Proc. Natl. Acad. Sci., 76, 5755.

Milo, G.E. and DiPaola, J.A. (1978). "Neoplastic transformation of human diploid cells *in vitro* after chemical carcinogen treatment," Nature 275, 130.

Milo, G.E., Oldham, J., Zimmerman, R., Hatch, G., and Weisbrode, S. (1981). "Characterizations of human cells transformed by chemical and physical carcinogens *in vitro*," In vitro 17, 719.

Mitelman, F. (1984). "Restricted number of chromosomal regions implicated in aetiology of human cancer and leukemia," Nature 310, 325.

Mix, M.C. and Schaffer, R.L. (1979). "Benzo[a]pyrene concentrations in mussels *(Mytilus edulis)* from Yaquina Bay, Oregon during June 1976–May 1978," Bull. Environ. Contam. Toxicol, 23, 677.

Modan, B., Ron, E., and Werner, A. (1977). "Thyroid cancer following scalp irradiation," Radiology 123, 741.

Moble, J., Whittemore, A., Pike, M., and Darby, S. (1985). "Gonadotrophins and ovarian cancer risk," J. Natl. Cancer Inst. 75, 178.

Mole, R.H. (1978). "The radiobiological significance of the studies with ^{224}Ra and Thorotrast," Health Phys. 35, 167.

Moloney, W.C. and Kastenbaum, M.R. (1955). "Leukemogenic effects of ionizing radiation on atomic bomb survivors in Hiroshima City," Science 121, 308.

Mondal, S. (1975). "Transformed liver cells obtained in culture from hepatectomized rats treated with dimethylnitrosamine (DMN)," Br. J. Cancer 31, 245.

MONSON, R.P. AND MACMAHON, B. (1984). "Pre-natal x-ray exposure and cancer in children," page 97 in *Radiation Carcinogenesis: Epidemiology and Biological Significance*, BOICE, J.D., JR. AND FRAUMENI, J.F., JR., Eds. (Raven Press, New York).

MONTESANO, R., SAFFIOTTI, U., AND RERRERO, A. (1974). "Synergistic effects of benzo[a]pyrene and diethylnitrosamine on respiratory carcinogenesis in hamsters," J. Natl. Cancer Inst. 53, 1395.

MONTESANO, R., SAFFIOTTI, U., AND SHUBIK, P. (1970). "The role of topical and systemic factors in experimental respiratory carcinogenesis," page 353 in *Inhalation Carcinogenesis*, AEC Symposium Series CONF 691001 (National Technical Information Service, Springfield, Virginia).

MONTESANO, R., SAINT VINCENT, L., AND TOMATIS, L. (1973). "Malignant transformation *in vitro* of rat liver cells by dimethylnitrosamine and N-methyl-N'-nitro-N-nitrosoguanidine," Br. J. Cancer 28, 215.

MOOLGAVKAR, S.H. AND KNUDSON, A.G. (1981). "Mutation cancer: a model for human carcinogenesis," J. Natl. Cancer Inst. 66, 1037.

MORDAN, L.J., MARTNER, J.E., AND BERTRAM, J.S. (1982). "Transformation frequency of C3H/10T1/2 cells depends on colony size of initiated cells at confluence," Proc. Am. Assoc. Cancer Res. 23, 73.

NAKAMURA, K., MCGREGOR, D.H., KATO, H., AND WAKABAYASHI, T. (1977). *Epidemiologic Study of Breast Cancer in A-bomb Survivors*, RERF TR 9-77 (Radiation Effects Research Foundation, Hiroshima).

NAKANISHI, Y. AND SCHNEIDER, E.L. (1979). "*In vivo* sister-chromatid exchange: A sensitive measure of DNA damage," Mutat. Res. 60, 329.

NAMBA, M., NISHITANI, K., AND KIMOTO, T. (1978). "Carcinogenesis in tissue culture. 29: Neoplastic transformation of a normal human diploid cell strain, WI-38, with Co-60 gamma rays," Jpn J. Exp. Med. 48, 303.

NARISAWA, T., MAGADIA, N.E., WEISBURGER, J.H., AND WYNDER, E.L. (1974). "Promoting effect of bile acids on colon carcinogenesis after intrarectal instillation of N-methyl-N'-nitro-N-nitrosoguanidine in rats," J. Natl. Cancer Inst. 55, 1093.

NAS/NRC (1956). National Academy of Sciences/National Research Council. Report of the Committee on the Biological Effects of Atomic Radiation, *The Biological Effects of Atomic Radiation, Pathological Effects of Atomic Radiation* (National Academy Press, Washington).

NAS/NRC (1972). National Academy of Sciences/National Research Council. *The Effects of Populations of Exposure to Low Levels of Ionizing Radiation*, Report of the Advisory Committee on the Biological Effects of Ionizing Radiation (BEIR) (National Academy Press, Washington).

NAS/NRC (1975). National Academy of Sciences/National Research Council. *Health Effects of Chemical Pesticides*, Report of the Consultative Panel on Health Hazards of Chemical Pesticides, Vol. 1 (National Academy Press, Washington).

NAS/NRC (1977). National Academy of Sciences/National Research Council. *Report on Drinking Water and Health* (National Academy Press, Washington).

NAS/NRC (1978). National Academy of Sciences/National Research Council. *Saccharin: Technical Assessment of Risks and Benefits, Part 1 of a 2-Part Study of* the Committee for a Study on Saccharin and Food Safety Policy, Panel I: Saccharin and Its Impurities (National Academy Press, Washington).

NAS/NRC (1980). National Academy of Sciences/National Research Council. *The Effects on Populations of Exposure to Low Levels of Ionizing Radiation*, Advisory Committee on the Biological Effects of Ionizing Radiation (BEIR) (National Academy Press, Washington).

NAS/NRC (1983a). National Academy of Sciences/National Research Council. *Identifying and Estimating the Genetic Impact of Chemical Mutagens* (National Academy Press, Washington).

NAS/NRC (1983b). National Academy of Sciences/National Research Council. *Risk Assessment in the Federal Government: Managing the Process* (National Academy Press, Washington).

NAS/NRC (1984). National Academy of Sciences/National Research Council. Steering Committee on Identification of Toxic and Potentially Toxic Chemicals for Considerations by the National Toxicology Program, *Toxicity Testing, Strategies to Determine Needs and Priorities* (National Academy Press, Washington).

NCRP (1954). National Council on Radiation Protection and Measurements. *Permissible Dose from External Sources of Ionizing Radiation* (1954) including *Maximum Permissible Exposure to Man, Addendum to National Bureau of Standards Handbook 59* (1958) (National Council on Radiation Protection and Measurements, Bethesda, Maryland).

NCRP (1980). National Council on Radiation Protection and Measurements. *Influence of Dose and Its Distribution in Time on Dose-Response Relationships for Low-LET Radiations*, NCRP Report No. 64 (National Council on Radiation Protection and Measurements, Bethesda, Maryland).

NCRP (1985). National Council on Radiation Protection and Measurements. *SI Units in Radiation Protection and Measurements*, NCRP Report No. 82 (National Council on Radiation Protection and Measurements, Bethesda, Maryland).

NCRP (1987a). National Council on Radiation Protection and Measurements. *Ionizing Radiation Exposure of the Population of the United States*, NCRP Report No. 93 (National Council on Radiation Protection and Measurements, Bethesda, Maryland).

NCRP (1987b). National Council on Radiation Protection and Measurements. *Recommendations on Limits for Exposure to Ionizing Radiation*, NCRP Report No. 91 (National Council on Radiation Protection and Measurements, Bethesda, Maryland).

NEBERT, D.W. (1981). "Genetic differences in susceptibility to chemically induced myelotoxicity and leukemia," Environ. Health Perspect. **39**, 11.

NEBERT, D.W. AND BAUSSERMAN, L.L. (1970). "Fate of inducer during induction of aryl hydrocarbon hydroxylase activity in mammalian cell culture. II. Levels of intracellular content of polycyclic hydrocarbon during enzyme induction and decay," Mol. Pharmacol. **6**, 304.

NEBERT, D.W. AND GONZALEZ, F.J. (1985). "Cytochrome P-450 gene expression and regulation," Trends in Pharmacol. Sci. **6**, 160.

NEBERT, D.W., JENSEN, N.M., LEVITT, R.C., AND FELTON, J.S. (1980). "Toxic chemical depression of the bone marrow and possible aplastic anemia explainable on a genetic basis," Clin. Toxicol. **16**, 99.

NEBERT, D.W., EISEN, H.J., NEGISHI, M., LANG, M.A., HJELMELAND, L.M., AND OKEY, A.B. (1981). "Genetic mechanisms controlling the induction of polysubstrate monooxygenase (P-450) activities," Ann. Rev. Pharmacol. Toxicol., **21**, 431.

NETTESHEIM, P. AND GRIESEMER, R.A. (1978). "Experimental models for studies of respiratory tract carcinogenesis," page 75 in *Pathogenesis and Therapy of Lung Cancer*, Harris, C.C., Ed. (Marcel Dekker, New York).

NEWBOLD, R.F. AND OVERELL, R.W. (1983). "Fibroblast immortality is a prerequisite for transformation by EJ c-Ha-ras oncogene," Nature **304**, 648.

NIGRO, N.D., SINGH, D.V., CAMPBELL, R.L., AND SOOK, M. (1975). "Effect of dietary beef fat on intestinal tumor formation by azoxymethane in rats," J. Natl. Cancer Inst. **54**, 439.

NISHIZUKA, Y., NAKAKUKI, K., AND SAKAKURA, T. (1964). "Induction of pulmonary tumors and leukemia by a single injection of 4-nitroquinoline 1-oxide to newborn and infant mice," Gann **55**, 495.

NRC (1984). Nuclear Regulatory Commission, *Code of Federal Regulations, Title 10, Par 50*, Appendix I (Government Printing Office, Washington).

ODASHIMA, S. (1969). "Development of leukemia in rats by oral administration of N-nitrosobutylurea in the drinking water," Gann **60**, 237.

ODASHIMA, S. (1970). "Leukemogenesis of N-nitrosobutylurea in the rat. I. Effect of various concentrations in the drinking water to female Donryu rats," Gann **61**, 245.

OGIU, T. (1978). "Hematological and histopathological characteristics of leukemias induced by 1-alkyl-1-nitrosourease in Donryu rats," Gann **69**, 781.

OKEY, A.B., BONDY, G.P., MASON, M.E., NEBERT, D.W., FORSTER-GIBSON, C., MUNCAN, J., AND DUFRESNE, M.J. (1980). "Temperature-dependent cytosol-to-nuclease translocation of the Ah receptor for 2,3,7,8-tetrachlorodi-benzo-p-dioxin in continuous cell culture lines," J. Biol. Chem. **255**, 11415.

O'NEILL, J.P. AND HSIE, A.W. (1979). "The CHO/HGPRT mutation assay: Experimental procedure," page 55 in *Banbury Report 2, Mammalian Cell Mutagenesis: The Maturation of Test Systems*, HSIE, A.W., O'NEILL, J.P., AND MCELHENY, V.K., Eds. (Cold Spring Harbor Laboratory Press, Cold Spring Harbor, New York).

OSHIMURA, M., HESTERBERG, T.W., TSUTSUI, T., AND BARRETT, J.C. (1984). "Correlation of asbestos-induced cytogenetic effects with cell transformation of Syrian hamster embryo cells in culture," Cancer Res. **44**, 5017.

OSTP (1985). Office of Science and Technology Policy. "Chemical Carcinogens: a review of the science and its associated principles," Federal Register 50 (50): 10371 (U.S. Government Printing Office, Washington).

PAI, S.B., STEELE, V.E., AND NETTESHEIM, P. (1983). "Quantitation of early cellular events during neoplastic transformation of primary tracheal epithelial cell cultures," Carcinogenesis **4**, 369.

PARK, H.Y. AND KOPROWSKA, I. (1968). "A comparative *in vivo* study of induced cervical lesions of mice," Cancer Res. **28**, 1478.

PARODI, S. AND BRAMBILLA, G. (1977). "Relationship between mutation and transformation frequencies in mammalian cells treated *in vitro* with chemical carcinogens," Mutat. Res. **47**, 53.

PATERSON, M.C., SMITH, P.J., BECH-HANSEN, N.T., SMITH, B.P., AND SELL, B.M. (1979). "Gamma-ray hypersensitivity and faulty DNA repair in cultured cells from humans exhibiting familial cancer proneness," page 484 in *Radiation Research, Proceedings of the 6th International Conference*, OKADA, S., IMAMURA, M., TERASIMA, T., AND YAMAGUCHI, H., Eds. (Japanese Association of Radiation Research, Tokyo).

PAWLOWSKI, A., HUBERMAN, E., AND MINON, I.A. (1980). "Skin melanoma induced by 7,12-dimethylbenzanthracene in albino guinea pigs and its similarities to skin melanoma of humans," Cancer Res. **40**, 3652.

PELKONEN, O. AND NEBERT, D.W. (1982). "Metabolism of polycyclic aromatic hydrocarbons: Etiologic role in carcinogenesis," Pharmacol. Rev. **34**, 189.

PELKONEN, O., BOOBIS, A.R., LEVITT, R.C., KOURI, R.E., AND NEBERT, D.W. (1979). "Genetic differences in the metabolic activation of benzo[a]pyrene in mice. Attempts to correlate tumorigenesis with binding or reactive intermediates to DNA and with mutagenesis *in vitro*," Pharmacology **18**, 281.

PENMAN, B.W., HOPPE, IV, H., AND THILLY, W.G. (1979). "Concentration-dependent mutation by alkylating agents in human lymphoblasts and *Salmonella typhimurium:* N-methyl-N-nitrosourethane and β-propiolactone," J. Natl. Cancer Inst. **63**, 903.

PERERA, F.P. (1981). "Carcinogenicity of airborne fine particulate benzo[a]pyrene: an appraisal of the evidence and the need for control," Environ. Health Perspect. **42**, 163.

PERERA, F.P., POIRIER, M.C., YUSPA, S.H., NAKAYAMA, J., JARETZKI, A., CURNEN, M.M., KNOWLES, D.M., AND WEINSTEIN, I.B. (1982). "A pilot project in molecular cancer epidemiology: determination of benzo[a]pyrene-DNA adducts in animal and human tissues by immunoassays," Carcinogenesis **3**, 1405.

PETO, R. (1977). "Epidemiology, multistage models, and short-term mutagenicity tests," page 1403 in *Origins of Human Cancer*, HIATT, H.H., WATSON, J.D., AND WINSTEN, J.A., Eds. (Cold Spring Harbor Laboratory Press, Cold Spring Harbor, New York).

PITOT, H.C., BARSNESS, L., GOLDSWORTHY, T., AND KITAGAWA, T. (1978). "Biochemical characterization of stages of hepatocarcinogenesis after a single dose of diethylnitrosamine," Nature **271**, 456.

POCHIN, E.E. (1974). "Occupational and other fatality rates," Community Health, **6**, 2.

POHL-RÜLING, J., FISCHER, P., HAAS, O., OBE, G., NATARAJAN, A.T., VAN-BUUL, P.P.W., BUCTON, K.E., BIANCHI, N.O., LARRAMENDY, M., KUCEROVA, M., POLIKOVA, Z., LEONARD, A., FABRY, L., PALITTI, F., SHARMA, T., BINDER, W., MUKHERJEE, R.N., AND MUKHERJEE, U. (1983). "Effect of low-dose acute x-irradiation on the frequencies of chromosomal aberrations in human peripheral lymphocytes in vitro," Mutat. Res. **110**, 71.

POLAND, A., TALEN, D., AND GLOVER, E. (1982). "Tumor promotion in mouse epidermis by 2,3,7,8- tetrachloridibenzo-p-diotin," Nature **300**, 271.

PONDER, B.A.J. (1980). "Genetics and Cancer," Biochim. Biophys. Acta **605**, 368.

POPESCU, N.C., AMSBAUGH, S.C., AND DiPAOLO, J.A. (1984). "Correlation of morphological transformation to sister chromatid exchanges induced by split doses of chemical or physical carcinogens on cultured Syrian hamster cells," Cancer Res. **44**, 1933.

POPPER, H., SELIKOFF, I.J., AND MALTONI, C. (1977). "Comparisons of neoplastic hepatic lesions in man and experimental animals," page 1359 in *Origins of Human Cancers.* HIATT, H.H., WATSON, J.D., AND WINSTEN, J.A., Eds. (Cold Spring Harbor Laboratory Press, Cold Spring Harbor, New York).

POZHARISSKI, K.M. (1975). "The significance of nonspecific injury for colon carcinogenesis in rats," Cancer Res. **35**, 3284.

PREHN, R.T. (1977). "Immunostimulation of the lymphodependent phase of neoplastic growth," J. Natl. Cancer Inst. **59**, 1043.

PRENTICE, R., YASHIMOTO, Y., AND MASON, M.W. (1983). "Relationship of cigarette smoking and radiation exposure to cancer mortality in Hiroshima and Nagasaki," J. Natl. Cancer Inst. **70**, 611.

PRESTON, R.J., AU, W., BENDER, M.A., BREWEN, J.G., CARRANO, A.V., HEDDLE, J.A., McFEE, A.F., WOLFF, S., AND WASSOM, J.S. (1983). "Mammalian *in vivo* and *in vitro* cytogenetic assays: A report of the Gene-Tox Program," Mutat. Res. **87**, 143.

PURCHASE, I.F.H. (1982). "An appraisal of predictive tests for carcinogenesis," Mutat. Res. **99**, 53.

RADFORD, E.P., AND ST. CLAIR RENARD, K.G. (1984). "Lung cancer in Swedish iron miners exposed to low doses of radon daughters," New Engl. J. Med. **310**, 1485.

RADMAN, M., JEGGO, P., AND WAGNER, R. (1982). "Chromosomal rearrangement and carcinogenesis," Mutat. Res. **98**, 249.

RALL, J.E., BEEBE, G.W., HOEL, D.G., JABLON, S., LAND, C.E., NYGAARD, O.F., UPTON, A.C., YALOW, R.S., AND ZEVE, V.H. (1985). *Report of the National Institutes of Health Working Group to Develop Radioepidemiological Tables,* NIH Pub. No. 85-2748. (Government Printing Office, Washington)

RANDERATH, E., AVITTS, T.A, REDDY, M.V., MILLER, R.H., EVERSON, R.B., AND RANDERATH, K. (1986). "Comparative ^{32}P-Analysis of cigarette smoke-induced DNA damage in human tissues and mouse skin," Cancer Res. **46**, 5869.

RASSOULZADEGAN, M., NAGHASHFAR, Z., COWIE, A., CARR, A., GRISONI, M., KAMEN, R., AND CUZIN, F. (1983). "Expression of the large T protein of polyoma virus promotes establishment in culture of "normal" rodent fibroblast cell lines," Proc. Natl. Acad. Sci. **80**, 4354.

REDDY, E.P., REYNOLDS, R.K., SANTOS, E., AND BARBACID, M. (1982). "A point mutation is responsible for the acquisition of transforming properties by the T24 human bladder carcinoma oncogene," Nature **300**, 149.

REES, E.D., MANDELSTAM, P., LOWRY, J.Q., AND LITSCOMB, H. (1971). "A study of the mechanisms of intestinal absorption of benzo[a]pyrene," Biochim. Biophys. Acta **225**, 96.

REGAN, J.D. AND SETLOW, R.B. (1973). "Repair of chemical damage of human

DNA," page 151 in *Chemical Mutagens: Principles and Methods for Their Detection*, Hollaender, A., Ed. (Plenum Press, New York).
RERF (1987). Radiation Effects Research Foundation. *U.S.-Japan Joint Reassessment of Atomic Bomb Radiation Dosimetry in Hiroshima and Nagasaki, Final Report*, Vol. I & II, Roesch, W.C., Ed. (Radiation Effects Research Foundation, Hiroshima, Japan).
REZNIKOFF, C.A., BERTRAM, J.S., BRANKOW, D.W., AND HEIDELBERGER, C. (1973) "Quantitative and qualitative studies of chemical transformation of cloned C3H mouse embryo cells sensitive to postconfluence inhibition of cell division," Cancer Res. 33, 3230.
RICHARDS, J., GUZMAN, R., YANG, J., NANDI, S., AND KONRAD, M. (1980). "Chemical carcinogenesis of mammary epithelium in cell culture," page 467 in *Cell Biology of Breast Cancer*, MCGRATH, C.M., BRENNAN, M.J., AND RICH, M.A., Eds. (Academic Press, New York).
RON, E. AND MODAN, B. (1984). "Thyroid and other neoplasms following childhood scalp irritation," page 139 in *Radiation Carcinogenesis: Epidemiology and Biological Significance*, BOICE, J.D., JR. AND FRAUMENI, J.F., JR., Eds. (Raven Press, New York).
ROSCOE, J.P. AND CLAISSE, P.J. (1978). "Analysis of N-ethyl-N-nitrosourea induced brain carcinogenesis by sequential culturing during the latent period. I. Morphology and tumorigenicity of the cultured cells and their growth in agar," J. Natl. Cancer Inst. 61, 381.
ROSEN, V.J. AND COLE, L.J. (1962). "Accelerated induction of kidney neoplasms in mice after x-irradiation (690 rad) and uninephrectomy," J. Natl. Cancer Inst. 28, 1031.
ROSSI, H.H. (1977). "The effects of small doses of ionizing radiation. Fundamental biophysical characteristics," Radiat. Res. 71, 1.
ROUTLEDGE, P.A. AND SHAND, D.G. (1979). "Presystemic drug elimination," Ann. Rev. Pharmacol. Toxicol. 19, 447.
ROWLAND, R.E., STEHNEY, A.F., AND LUCAS, H.F., JR. (1978). "Dose-response relationships for female radium dial workers," Radiat. Res. 76, 368.
ROWLEY, J.D. (1982). "Identification of the constant chromosome regions involved in human hematological malignant disease," Science 216, 749.
ROWLEY, J.D. (1984). "Biological implications of consistent chromosome rearrangements in leukemia and lymphoma," Cancer Res. 44, 3159.
RULEY, H.E. (1983). "Adenovirus early region 1A enables viral and cellular transforming genes to transform primary cells in culture," Nature 304, 602.
SACCOMANO, G., YALE, C., DIXON, W., AUERBACH, O., AND HUTH, G. (1986). "An epidemiological analysis of the relationship between exposure to Rn progeny, smoking and bronchogenic carcinoma in the Uranium mining population of the Colorado plateau - 1960–1980," Health Phys. 50, 605.
SAFFIOTTI, U. (1973). "Metabolic host factors in carcinogenesis," page 243 in *Host Environment Interactions in the Etiology of Cancer in Man*. DOLL, R., VODOPIJA, I. AND DAVIS, W., Eds. (International Agency for Research on Cancer, Lyons).
SAFFIOTTI, U. (1975). "Symposium: Current progress in pancreatic carcinogenesis research," Cancer Res. 35, 2223.

SANDBERG, A.A., Ed. (1982). *Sister Chromatid Exchange, Program and Topics in Cytogenetics*, Vol. 2., (Alan R. Liss, Inc., New York).
SANDBERG, A.A., (1983). "A Chromosomal hypothesis of oncogenesis," Cancer Genet. Cytogenet. 8, 277.
SANTODONATO, J., HOWARD, P., AND BASU, D. (1981). "Health and ecological assessment of polynuclear aromatic hydrocarbons," J. Environ. Pathol. and Toxicol. 5, 1.
SASAKI, S., TSUTOMU, K., SATO, P., AND KAWASHIMA, N. (1978). "Late effects of fetal mice irradiated at middle or late uterine stage," Gann 609, 167.
SATO, T. AND KAUFFMAN, S.L. (1980). "A scanning electron microscopic study of the type II and Clara cell adenoma of the mouse lung," Lab. Invest. 43, 28.
SAVAGE, J.R.K. (1975). "Classifications and relationships of induced chromosomal structural changes," J. Med. Genet. 12, 103.
SCHAEFFER, W.I. AND HEINTZ, N.H. (1978). "A diploid rat liver cell cultured," In Vitro 14, 418.
SCHMAEHL, D. (1970). "Experimental untersuchungen zur syncarcinogenese VI. Mitteilung: addition minimaler dosen von vier verschiendenen hepatotropen carcinogen bei der leberkrebserzeugung bei ratten," Z. Krebsforsch 74. 457.
SCHOTTENFELD, D. AND HAAS, J.F. (1978). "The workplace as a cause of cancer," Clin. Bull. 8, 54.
SELIKOFF, I.J. (1977). "Cancer risk of asbestos exposure," page 1765 in *Origins of Human Cancer*, Book C, HIATT, H.H., WATSON, J.D., AND WINSTEN, J.A., Eds. (Cold Spring Harbor Laboratory Press, Cold Spring Harbor, New York).
SHABAD, L.M., PYLEV, L.N., KRIVOSHEEVA, L.V., KULAGINA, T.F., AND NEMENKO, B.A. (1974). "Experimental studies on asbestos carcinogenicity," J. Natl. Cancer Inst. 52, 1175,
SHAIN, S.A., MCCULLOUGH, B., AND NITCHUK, W.M. (1979). "Primary and transplantable adenocarcinomas of the AxC rat ventral prostate: morphologic characterization and examination of C_{19}-steroid metabolism by early passage tumors," J. Natl. Cancer Inst. 62, 313.
SHAMSUDDIN, A.K., SINOPOLI, N.T., HEMMINKI, K., BOESCH, R.R., AND HARR, C.C. (1985). "Detection of benzo[a]pyrene: DNA adducts in human white blood cells," Cancer Res. 45, 66.
SHAW, E.I. AND HSIE, A.W. (1978). "Conditions necessary for quantifying ethyl methanesulfonate-induced mutations to purine-analogue resistance in Chinese hamster V79 cells," Mutat. Res. 51, 237.
SHELLABARGER, C.J. (1981). "Pituitary and steroid hormones in radiation-induced mammary tumors," page 339 in *Hormones and Breast Cancer* Banbury Report 8, PIKE, M.C., SUTERI, P.K., AND WELCH, C.W., Eds. (Cold Spring Harbor Laboratory Press, Cold Spring Harbor, New York).
SHELLABARGER, C.J., BOND, V.P., CRONKITE, E.P., AND APONTE, G.E. (1969). "Relationship of dose to total-body ^{60}Co radiation to incidence of mammary neoplasia in female rats," page 161 in *Radiation-Induced Cancer*, IAEA/STI/PUB/228 (International Atomic Energy Agency, Vienna).
SHORE, R.E., HEMPLEMANN, L.H., KOWALUK, E., MANSUR, P.S., PASTERNACK, B.S., ALBERT, R.E., AND HAUGIE, G.E. (1977). "Breast neoplasms in women

treated with x rays for acute postpartum mastitis," J. Natl. Cancer Inst. **59**, 813.

SHORE, R.E., WOODARD, E.D., HEMPLEMANN, L.H., AND PASTERNACK, B.S. (1980). "Synergism between radiation and other risk factors for breast cancer," Prev. Med. **9**, 815.

SHORE, R.E., ALBERT, R.E., REED, M., HARLEY, N., AND PASTERNACK, B.S. (1984). "Skin cancer incidence among children irradiated for ringworm of the scalp," Radiat. Res. **110**, 192.

SHORE, R.E., WOODARD, E., HILDRETH, N., DVORETSKY, P., HEMPLEMANN, L. AND PASTERNACK, P. (1985). "Thyroid tumors following thymus irradiation," J. Natl. Cancer Inst. **74**, 1177.

SHORE, R.E., HILDRETH, N., WOODARD, E., DVORETSKY, P., HEMPLEMANN, L., AND PASTERNACK, P. (1986). "Breast cancer among women given X-ray therapy for acute postpartum mastitis," J. Natl. Cancer Inst. **77**, 689.

SILINSKAS, K.C., KATELEY, S.A., TOWER, J.E., MAHER, V.M., AND MCCORMICK, J.J. (1981). "Induction of anchorage-independent growth in human fibroblasts by propane sultone," Cancer Res. **41**, 1620.

SINCLAIR, W.K. (1981). "The scientific basis for risk quantification," page 3 in *Quantitative Risk in Standards Setting*, Proceedings of the Sixteenth Annual Meeting of the National Council on Radiation Protection and Measurements, Proceedings No. 2 (National Council on Radiation Protection and Measurements, Bethesda, Maryland).

SINCLAIR, W.K. (1983). "Fifty years of neutrons in biology and medicine: The comparative effects of neutrons in biological systems," page 1 in: Proc. Eighth Symposium in Microdosimetry, BOOZ, J. AND EBERT, H., Eds. EUR 8395 (Commission European Communities, Luxembourg).

SINGER, B. AND GRUNBERGER, D. (1983). *Molecular Biology of Mutagens and Carcinogens* (Plenum Press, New York).

SINGER, B. AND KUSMIEREK, J.T. (1982). "Chemical mutagenesis," Ann. Rev. Biochem. **51**, 655.

SIVAK, A. AND VAN DUUREN, B. (1967). "Induction in cell culture by a phorbol ester," Science **15**, 1443.

SIZARET, P., MALAVEILLE, C., MONTESANNO, R., AND FRAYSSINET, C. (1982). "Detection of aflatoxins and related metabolites by radioimmunoassay," J. Natl. Cancer Inst. **69**, 1375.

SLAGA, T.J. AND BOUTWELL, R.K. (1977). "Inhibition of the tumor-initiating ability of the potent carcinogen 7,12-dimethylbenz[a]anthracene by the weak tumor initiator 1,2,3,4-dibenzanthracene," Cancer Res. **37**, 128.

SMITH, P.G. AND DOLL, R. (1982). "Mortality among patients with ankylosing spondylitis after a single treatment course with x rays," Br. Med. J. **284**, 449.

SNYDER, C.A., GOLDSTEIN, B.D., SELLAKUMAR, A., BROMBERG, I., LASKIN, S., AND ALBERT, R.E. (1980). "The inhalation toxicity of benzene: incidence of hematopoietic neoplasm and hematotoxicity in AKR/J and C57BL/6J mice," Toxicol. Appl. Pharmacol. **54**, 323.

SPIEGELHALDER, B. AND PREUSSMANN, R. (1983). "Occupational nitrosamine exposure. 1. Rubber and tyre industry," Carcinogenesis **4**, 1147.

SPIRTAS, R., BEEVE, G., BAXTER, P., DACEY, E., FABER, M., FALK, H., VAN

KAICK, G. AND STAFFORD, J. (1983). "Angiosarcoma as a model for comparative carcinogenesis," Lancet 2, 456.
SPORN, M.B. (1977). "Retinoids and carcinogenesis," Nutrition Rev. 35, 65.
STEELE, V.E., MARCHOK, A.C., AND NETTENSHEIM, P. (1979). "Oncogenic transformation in epithelial cell lines derived from tracheal explants exposed in vitro to N-methyl-N-nitro'N'nitrosoguanine," Cancer Res. 39, 3805.
STEGLICH, C.S. AND DEMARS, R. (1982). "Mutations causing deficiency of APRT in fibroblasts cultured from humans heterozygous for mutant APRT alleles," Somatic Cell Genet. 8, 115.
STETKA, D.G. AND WOLFF, S. (1976a). "Sister chromatid exchange as an assay for genetic damage induced by mutagen-carcinogens. I. In vivo test for compounds requiring metabolic activation," Mutat. Res. 41, 333.
STETKA, D.G. AND WOLFF, S. (1976b). "Sister chromatid exchange as an assay for genetic damage induced by mutagen-carcinogens. II. In vitro test for compounds requiring metabolic activation," Mutat. Res. 41, 343.
STEWART, A. (1960). "Leukemia and prenatal x rays," Br. Med. J. 2, 1381.
STOCKLIN, G. (1979). "Chemical and biological effects of β-decay and inner shell ionization in biomolecules: a new approach to radiation biology," page 382 in Radiation Research, OKADA, S., IMAMURA, M., TERASIMA, T., AND YAMAGUCHI, H., Eds. (Toppan Printing Company, Tokyo).
STRAUSS, G.H. AND ALBERTINI, R.J. (1979). "Enumeration of 6-thioguanine-resistant peripheral blood lymphocytes in man as a potential test for somatic cell mutations arising in vivo,"Mutat. Res. 61, 353.
STYLES, J.A. (1980). "Studies on the detection of carcinogens using a mammalian cell transformation assay with liver homogenate activation," page 226 in Short-Term Test Systems for Detecting Carcinogens, NORPOTH, K.H. AND GARNER, R.C., Eds. (Springer-Verlag, Berlin).
SUKUMAR, S., NOTARIO, V., MARTIN-ZANCA, D., AND BARBACID, M. (1983). "Induction of mammary carcinomas in rats by nitroso-methylurea involves malignant activation of H-ras-1 locus by single point mutations," Nature 306, 658.
SUMMERHAYES, I.C. (1979). "Influence of donor age on In vitro transformation of bladder epithelium," page 137 in Neoplastic Transformation in Different Epithelial Cell Systems in Vitro, FRANKS, L.M. AND WIGLEY, C.B., Eds. (Academic Press London).
SUMMERHAYES, I.C., CHENG, Y.E., SUN, T.T., AND CHEN, L.B. (1981). "Expression of keratin and vimentin intermediate filaments in rabbit bladder epithelial cells at different stages of benzo[a]pyrene-induced neoplastic progression," J. Cell Bio. 90, 63.
SUTHERLAND, B.M., CIMINO, V.S., DELIHAS, M., SHIH, A.G., AND OLIVER, R.P. (1980). "Ultraviolet light-induced transformation of human cells to anchorage-independent growth," Cancer Res. 40, 1934.
SYDNOR, K.L., BUTENANDT, O., BRILLANTES, F.P., AND HUGGINS, C. (1962). "Race-strain factor related to hydrocarbon-induced mammary cancer in rats," J. Natl. Cancer Inst. 29, 805.
TABIN, C.J., BRADLEY, S.M., BARGMANN, C.I., WEINBERG, R.A., PAPAGEORGE, A.G., SCOLNICK, E.M., DHAR, R., LOWY, D.R., AND CHANGE, E.H.

(1982). "Mechanism of activation of a human oncogene," Nature 300, 143.
TAKAHASHI, M., POUR, P., ALTHOFF, J., AND DONNELLY, T. (1977). "Sequential alteration of the pancreas during carcinogenesis in Syrian hamsters by N-nitrosobis(2-oxolpropyl)amine," Cancer Res. 37, 4602.
TAKAI, Y., KISHIMOTO, A., AND NISHIZUKA, Y. (1982). "Calcium and phospholipid turnover as transmembrane signaling for protein phosphorylation," pages 386 in *Calcium and Cell Function*, Vol. II, CHEUNG, W.Y., Ed. (Academic Press, New York).
TEEBOR, G.W. AND FRENKEL, K. (1983). "The initiation of DNA excision-repair," Adv. Cancer Res. 38, 23.
TERZAGHI, M. AND LITTLE, J.B. (1976a). "Oncogenic transformation after split dose x-irradiation," Int. J. Radiat. Biol. 29, 583.
TERZAGHI, M. AND LITTLE, J.B. (1976b). "X-radiation-induced transformation in a C3H mouse embryo-derived cell line," Cancer Res. 36, 1367.
TERZAGHI, M. AND NETTESHEIM, P. (1979). "Dynamics of neoplastic development in carcinogen-exposed tracheal mucosa," Cancer Res. 39, 4003.
TESTA, B. AND JENNER, P. (1978). "Novel drug metabolites produced by functionalization reactions: chemistry and toxicology," Drug Metab. Rev. 7, 325.
THIELMANN, H.W., SCHRODER, C.H., O'NEILL, J.P., BRIMER, P.A., AND HSIE, A.W. (1979). "Relationship between DNA alkylation and specific-locus mutation induced by N-methyl- and N-ethyl-N-nitrosourea in cultured Chinese hamster ovary cells (CHO/HGPRT system)," Chem. Biol. Interact. 26, 233.
THILLY, W.G. (1979). "Study of mutagenesis in diploid human lymphoblasts," page 341 in *Banbury Report 2, Mammalian Cell Mutagenesis: The Maturation of Test Systems,* HSIE, A.W., O'NEILL, J.P. AND MCELHENY, V.K., Eds. (Cold Spring Harbor Laboratory Press, Cold Spring Harbor, New York).
THILLY, W.G., DELUCA, J.G., FURTH, E.E., HOPPE, H. IV, KADEN, D.A., KROLEWSKI, J.J., LIBER, H.L., SKOPEK, T.R., SLAPIKOFF, S.A., TIZZARD, R.J. AND PENMAN, B.W. (1980). "Gene-locus mutation assays in diploid human lymphoblast lines," page 331 in *Chemical Mutagens*, Vol. 6, DESERRES, F.J. AND HOLLAENDER, A., Eds. (Plenum Press, New York).
THOMASSEN, D.G., GRAY, T., MASS, M.J., AND BARRETT, J.C. (1983). "High frequency of carcinogen-induced early preneoplastic changes in rat tracheal epithelial cells in culture," Cancer Res. 43, 5956.
THOMPSON, L.H., BROOKMAN, K.W., CARRANO, A.V., AND DILLEHAY, L.E. (1982). "Role of DNA repair in mutagenesis of Chinese hamster ovary cells by 7-bromomethylbenz[a]anthracene," Proc. Natl. Acad. Sci. 79, 534.
THOMSON, J.F., LOMBARD, L.S., GRAHN, D., WILLIAMSON, F.S. AND FRITZ, T.F. (1982). "RBE of fission neutrons for life shortening and tumorigenesis," page 75 in *Neutron Carcinogenesis*, BROERSE, J.J. AND GERBER, G.B., Eds. (Commission of the European Communities, Luxembourg).
THORGEIRSSON, S.S. AND NEBERT, D.W. (1977). "The *Ah* locus and the metabolism of chemical carcinogens and other foreign compounds," Adv. Cancer Res. 25, 149.
THURNHERR, N., DESCHNER, E.E., STONEHILL, E.H., AND LIPKIN, M. (1973). "Induction of adenocarcinomas of the colon in mice by weekly injections of 1, 2-dimethylhydrazine," Cancer Res. 33, 940.

TICE, R.R. (1984). "An overview of occupation studies directed at assessing genetic damage." page 439 in *Reproduction: The New Frontier in Occupational and Environmental Health Research*, LOCKEY, J.E., LEMASTERS, G.K., AND KOYE, W.R., JR. Eds. (Alan R. Liss, Inc., New York).

TICE, R.R. AND HOLLAENDER, A. (1984a). *Sister Chromatid Exchange: Twenty-Five Years of Experimental Research, V.A.: The Nature of SCEs*, TICE, R.R. AND HOLLAENDER, A., Eds. (Plenum Press, New York).

TICE, R.R. AND HOLLAENDER, A. (1984b). *Sister Chromatid Exchange: Twenty-Five Years of Experimental Research, V.B.: Genetic Toxicology and Human Studies*, TICE, R.R. AND HOLLAENDER, A., Eds. (Plenum Press, New York).

TOKUNAGA, M., NORMAN, J.E., JR., ASANO, M., TOKUOKA, S., EZAKI, H., NISHIMORI, I., AND TSUJI, Y. (1979). "Malignant breast tumors among atomic bomb survivors, Hiroshima and Nagasaki, 1950-1974," J. Natl. Cancer Inst. **62**, 1347.

TOKUNAGA, M., LAND, C.E., ASANO, M., TOKUOKA, S., EZAKI, H., NISHIMORI, I., AND FUJIKURA, T. (1984). "Breast cancer among atomic bomb survivors," Page 45 in *Radiation Carcinogenesis: Epidemiology and Biological Significance*, BOICE, J.D., JR. AND FRAUMENI, J.F., JR., Eds. (Raven Press, New York).

TOMATIS, L. (1965). "Increased incidence of tumors in F1 and F2 generations from pregnant mice injected with a polycyclic hydrocarbon," Proc. Soc. Exp. Biol. Med. **119**, 743.

TOMATIS, L. AND MOHR, U. (1973). *Transplacental Carcinogenesis*, (International Agency for Research on Cancer, Lyon).

TOMATIS, L., HILFRICH, J., AND TUROSOV, V. (1975). "The occurrence of tumors in F_1, F_2 and F_3 descendents of BD rats exposed to N-nitrosomethylurea during pregnancy," Int. J. Cancer **15**, 385.

TOMATIS, L., AGATHA, C., BARTSCH, H., HUFF, J., MONTESANO, R., SARACCI, R., WALKER, E., AND WILBOURN, J. (1978). "Evaluation of the carcinogenicity of chemicals: A review of the monograph program of the International Agency for Research on Cancer," Cancer Res. **38**, 877.

TOTH, B. (1975). "Synthetic and naturally occurring hydrazines as possible cancer causative agents." Cancer Res. **35**, 3693.

TUCKER, M.A., BOICE, J.D., JR., HOOVER, R.N. AND MEADOWS, A.T. (1984). "Cancer risk following treatment of childhood cancer," page 211 in *Radiation Carcinogenesis: Epidemiology and Biological Significance*, BOICE, J.D., JR., AND FRAUMENI, J.F., JR., Eds. (Raven Press, New York).

ULRICH, H. (1946). "The incidence of leukemia in radiologists," New Engl. J. Med. **234**, 45.

ULLRICH, R.L., JERNIGAN, M.C., COSGROVE, G.E., SATTERFIELD, L.C., BOWLES, N.D. AND STORER, J.B., (1976). "The influence of of dose and dose rate on the incidence of neoplastic disease in RFM mice after neutron irradiation," Radiat. Res. **68**, 115.

ULLRICH, R.L. AND STORER, J.B. (1978). "Influence of dose, dose-rate and radiation quality on radiation carcinogenesis and life shortening in RFM and BALB/C mice," in *Late Biological Effects of Ionizing Radiation*, Vol. II., Proceedings of a Symposium, Vienna, 1978, (International Atomic Energy Agency, Vienna).

UMEDA, M. AND IYPE, P.T. (1973). "An improved expression of *in vitro* transformation rate based on cytotoxicity produced by chemical carcinogens," Br. J. Cancer 28, 71.
UNSCEAR (1958). United Nations Scientific Committee on the Effects of Atomic Radiation. *Report to the General Assembly, Official Records: Thirteenth Session*, Supplement No. 17 (A/3838) (United Nations, New York).
UNSCEAR (1972). United Nations Scientific Committee on the Effects of Atomic Radiation. *Ionizing Radiation: Levels and Effects, Report to the General Assembly with Annexes* (United Nations, New York).
UNSCEAR (1977). United Nations Scientific Committee on the Effects of Atomic Radiation. *Sources and Effects of Ionizing Radiation, Report to the General Assembly with Annexes* (United Nations, New York).
UNSCEAR (1982). United Nations Scientific Committee on the Effects of Atomic Radiation. *Ionizing Radiation: Sources and Biological Effects, Report to the General Assembly with Annexes* (United Nations, New York).
UNSCEAR (1986). United Nations Scientific Committee on the Effects of Atomic Radiation. *Genetic and Somatic Effects of Ionizing Radiation, Report to the General Assembly with Annexes* (United Nations, New York).
UPTON, A.C. (1967). "Comparative observations on radiation carcinogenesis in man and animals," page 631 in *Carcinogenesis: A Broad Critique*. (Williams and Wilkins, Baltimore).
UPTON, A.C. (1977). "Radiobiological effects of low doses: implications for radiological protection," Radiat. Res., 71, 51.
UPTON, A.C. (1982). "Principles of cancer biology: Etiology and prevention," page 33 in *Principles and Practices of Oncology*, DEVITA, V.T., HELLMAN, S., AND ROSENBERG, S.A., Eds. (J.B. Lippincott Company, Philadelphia).
UPTON, A.C. (1984). "Biological aspects of radiation carcinogenesis," page 9 in *Radiation Carcinogenesis: Epidemiology and Biological Significance*, BOICE, J.D., JR. AND FRAUMENI, J.F., JR., Eds. (Raven Press, New York).
UPTON, A.C. AND FURTH, J. (1954). "The effects of cortisone on the development of spontaneous leukemia in mice and on its induction by irradiation," Blood 9, 686.
UPTON, A.C., WOLFF, F.F., FURTH, J., AND KIMBALL, A.W. (1958). "A comparison of the induction of myeloid and lymphoid leukemias in X-irradiated RF mice," Cancer Res. 18, 842.
UPTON, A.C., WOLFF, F.F., AND SNIFFEN, E.P. (1961). "Leukemogenic effect of myleran on the mouse thymus," Proceedings of the Society for Experimental Biology and Medicine 108, 464.
UPTON, A.C., CLAYSON, D.G., JANSEN, D.J., ROSENKRANZ, H., AND WILLIAMS, G. (1984). "Report of ICPEMC Task Group on the differentiation between Genotoxic and Non-Genotoxic Carcinogens," Mutat. Res., 133, 1.
VAN DUUREN, B.L. (1976). "Tumor-promoting and co-carcinogenic agents in chemical carcinogenesis," page 24 in *Chemical Carcinogens*, SEARLE, C.E., Ed. (American Chemical Society, Washington).
VAN DUUREN, B.L., SIVAK, A. AND SEIDMAN, I. (1975). "The effect of aging and interval between primary and secondary treatment in two-stage carcinogenesis on mouse skin," Cancer Res. 35, 502.
VAN ZEELAND, A.A. (1978). "Post-treatment with caffeine and the induction of

gene mutations by ultraviolet irradiation and ethyl methanesulfonate in V-79 Chinese hamster cells in culture," Mutat. Res. 50, 145.
WAKABAYASHI, T., KATO, H., IKEDA, R., AND SCHULL, W.J. (1983). "Incidence of cancer in 1950-1978 based on the Tumor Registry, Nagasaki," Radiat. Res. 93, 112.
WATTENBERG, L.W. (1978). "Inhibitors of chemical carcinogens," Adv. Cancer Res. 29, 197.
WEINBERG, R.A. (1982). "Oncogenes of spontaneous and chemically induced tumors," Adv. in Cancer Res. 36, 149.
WEISBURGER, J.H., HADIDIAN, Z., FREDRICKSON, T.N., AND WEISBURGER, E.K. (1967). "Host properties determine target, bladder or liver, in chemical carcinogenesis," page 45 in *Bladder Cancer*, DEICHMANN, W.B., Ed. (Aesculapius Publishing Co., Alabama).
WEISBURGER, J.H. AND WILLIAMS, G.M. (1975). "Metabolism of chemical carcinogens," page 185 in *Cancer: A Comprehensive Treatise* Vol. I., BECKER, F.F., Ed. (Plenum Press, New York).
WEISBURGER, J.H., REDDY, B.S., AND WYNER, E.L. (1977). "Colon Cancer: its epidemiology and experimental production," Cancer 40, 2414.
WELSCH, C.W. AND NAGASAWA, H. (1977). "Prolactin and murine mammary tumorigenesis: A review," Cancer Res. 37, 951.
WHITTEMORE, A.S. (1978). "Quantitative theories of oncogenesis," Adv. Cancer Res. 27, 55.
WHITTEMORE, A.S. AND MCMILLAN, A. (1983). "Lung cancer mortality among U.S. uranium miners: A reappraisal," J. Natl. Cancer Inst. 71, 489.
WIGLEY, C.B. (1979). "Transformation *in vitro* of adult mouse salivary gland epithelium; a system for studies on mechanisms of initiation and promotion," page 3 in *Neoplastic Transformation in Differentiated Epithelial Cell Systems In Vitro*, FRANKS, L.M. AND WIGLEY, C.B., Eds. (Academic Press, London).
WILLIAMS, G. (1976). "Functional markers and growth behavior of preneoplastic hepatocytes," Cancer Res. 36, 2540.
WILLIAMS, G.M., ELLIOT, J.M., AND WEISBURGER, J.H. (1973). "Carcinoma after malignant conversion *in vitro* of epithelial-like cells from rat liver following exposure to chemical carcinogens," Cancer Res. 33, 606.
WILLIAMS, M.H.C. (1958). "Occupational tumors of the bladder," page 337 in Cancer, RAVEN, R.W. Ed. (Butterworth, London).
WOGAN, G.N. (1976). "The induction of liver cell cancer by chemicals," page 121 in *Liver Cell Cancer*, CAMERON, H.M., LINSELL, D.A., AND WARWICK, G.P., Eds. (Elsevier, Amsterdam).
WOLFF, S., Ed. (1982). *Sister Chromatid Exchange*, (John Wiley and Sons, Inc. New York).
WOODARD, E.D. (1981). "Risk of thyroid cancer after irradiation in childhood," page 199 in *Cancer: Achievements, Challenges, and Prospects for the 1980s, Vol. 1.*, BURCHENAL, J.H. AND OETTGEN, H.F., Eds. (Grune and Stratton, New York).
WYNDER, E.L. AND HECHT, S. (1976). "Animal models," page 81 in *Lung Cancer*. UICC Technical Report Series, Vol. 25 (Imprimerie Montfort, Geneva).

YARITA, T., NETTESHEIM, P. AND WILLIAMS, M.L. (1978). "Tumor induction in the trachea of hamsters with N-nitroso-N-methylurea," Cancer Res. **38**, 1667.

YOKORO, K., IMAMURA, N., TAKIZAWA, S., NISHIHARA, H., AND NISHIHARA, E. (1970). "Leukemogenic and mammary tumorigenic effects of N-nitrosobutylurea in mice and rats," Gann **61**, 287.

YUASA, Y., SRIVASTAVA, S.K., DUNN, C.Y., RHIM, J.S., REDDY, E.P., AND AARONSON, S.A. (1983). "Acquisition of transforming properties by alternative point mutations within c-bas/has human proto-oncogene," Nature **303**, 775.

YUNIS, J.J. (1983). "The chromosomal basis of human neoplasia," Science **221**, 227.

YUSPA, S.H. (1983). "Cutaneous carcinogenesis, natural and experimental" page 1115 in *Biochemistry and Physiology of the Skin*, GOLDSMITH, L.A., Ed. (Oxford University Press, New York).

YUSPA, S.H. AND MORGAN, D.L. (1981). "Mouse skin cells resistant to terminal differentiation associated with initiation of carcinogenesis," Nature **293**, 72.

ZEISE, L., WILSON, R., AND CROUCH, E.A.C. (1987). "The dose response relationships for carcinogens: a review," Environ. Health Perspective **73**, 259.

The NCRP

The National Council on Radiation Protection and Measurements is a nonprofit corporation chartered by Congress in 1964 to:
1. Collect, analyze, develop, and disseminate in the public interest information and recommendations about (a) protection against radiation and (b) radiation measurements, quantities, and units, particularly those concerned with radiation protection;
2. Provide a means by which organizations concerned with the scientific and related aspects of radiation protection and of radiation quantities, units, and measurements may cooperate for effective utilization of their combined resources, and to stimulate the work of such organizations;
3. Develop basic concepts about radiation quantities, units, and measurements, about the application of these concepts, and about radiation protection;
4. Cooperate with the International Commission on Radiological Protection, the International Commission on Radiation Units and Measurements, and other national and international organizations, governmental and private, concerned with radiation quantities, units, and measurements and with radiation protection.

The Council is the successor to the unincorporated association of scientists known as the National Committee on Radiation Protection and Measurements and was formed to carry on the work begun by the Committee.

The Council is made up of the members and the participants who serve on the over sixty scientific committees of the Council. The scientific committees, composed of experts having detailed knowledge and competence in the particular area of the committee's interest, draft proposed recommendations. These are then submitted to the full membership of the Council for careful review and approval before being published.

The following comprise the current officers and membership of the Council:

Officers

President	WARREN K. SINCLAIR
Vice President	S. JAMES ADELSTEIN
Secretary and Treasurer	W. ROGER NEY
Assistant Secretary	CARL D. HOBELMAN
Assistant Treasurer	JAMES F. BERG

Members

SEYMOUR ABRAHAMSON
S. JAMES ADELSTEIN
PETER R. ALMOND
EDWARD L. ALPEN
JOHN A. AUXIER
WILLIAM J. BAIR
MICHAEL A. BENDER
BRUCE B. BOECKER
JOHN D. BOICE, JR.
ROBERT L. BRENT
ANTONE L. BROOKS
THOMAS F. BUDINGER
MELVIN W. CARTER
RANDALL S. CASWELL
JAMES E. CLEAVER
FRED T. CROSS
STANLEY B. CURTIS
GERALD D. DODD
PATRICIA W. DURBIN
JOE A. ELDER
THOMAS S. ELY
JACOB I. FABRIKANT
R. J. MICHAEL FRY
ETHEL S. GILBERT
ROBERT A. GOEPP
JOEL E. GRAY
ARTHUR W. GUY
ERIC J. HALL
NAOMI H. HARLEY
WILLIAM R. HENDEE
DONALD G. JACOBS
A. EVERETTE JAMES, JR.
BERND KAHN
KENNETH R. KASE
CHARLES E. LAND
GEORGE R. LEOPOLD
RAY D. LLOYD
ARTHUR C. LUCAS
CHARLES W. MAYS
ROGER O. MCCLELLAN
JAMES E. MCLAUGHLIN
BARBARA J. MCNEIL
THOMAS F. MEANEY
CHARLES B. MEINHOLD
MORTIMER L. MENDELSOHN
FRED A. METTLER
WILLIAM A. MILLS
DADE W. MOELLER
A. ALAN MOGHISSI
MARY ELLEN O'CONNOR
ANDREW K. POZNANSKI
NORMAN C. RASMUSSEN
WILLIAM C. REINIG
CHESTER R. RICHMOND
JAMES S. ROBERTSON
MARVIN ROSENSTEIN
LAWRENCE N. ROTHENBERG
LEONARD A. SAGAN
ROBERT A. SCHLENKER
WILLIAM J. SCHULL
ROY E. SHORE
WARREN K. SINCLAIR
PAUL SLOVIC
LEWIS V. SPENCER
WILLIAM L. TEMPLETON
THOMAS S. TENFORDE
J.W. THIESSEN
JOHN E. TILL
ROBERT ULLRICH
ARTHUR C. UPTON
GEORGE L. VOELZ
EDWARD W. WEBSTER
GEORGE M. WILKENING
H. RODNEY WITHERS
MARVIN ZISKIN

Honorary Members

LAURISTON S. TAYLOR, *Honorary President*

VICTOR P. BOND
REYNOLD F. BROWN
AUSTIN M. BRUES
GEORGE W. CASARETT
FREDERICK P. COWAN
JAMES F. CROW
MERRILL EISENBUD
ROBLEY D. EVANS
RICHARD F. FOSTER
HYMER L. FRIEDELL
ROBERT O. GORSON
JOHN H. HARLEY
JOHN W. HEALY
LOUIS H. HEMPELMANN, JR.
PAUL C. HODGES
GEORGE V. LEROY
WILFRID B. MANN
KARL Z. MORGAN
ROBERT J. NELSEN
WESLEY L. NYBORG
HARALD H. ROSSI
WILLIAM L. RUSSELL
JOHN H. RUST
EUGENE L. SAENGER
J. NEWELL STANNARD
ROY C. THOMPSON
HAROLD O. WYCKOFF

Currently, the following subgroups are actively engaged in formulating recommendations:

SC 1: Basic Radiation Protection Criteria
    SC 1-1 Probability of Causation for Genetic and Developmental Effects
    SC 1-2 Risk Estimates for Radiation Protection
SC 3: Medical X-Ray, Electron Beam and Gamma-Ray Protection for Energies Up to 50 MeV—Equipment Performance and Use
SC 16: X-Ray Protection in Dental Offices
SC 18B: Standards and Measurement of Radioactivity for Radiological Use—Medical and Biological Applications
SC 40: Biological Aspects of Radiation Protection Criteria
    SC 40-1 Atomic Bomb Survivor Dosimetry
    SC 40-1A Biological Aspects of Dosimetry of Atomic Bomb Survivors
SC 44: Radiation Associated with Medical Examinations
SC 45: Radiation Received by Radiation Employees
SC 46: Operational Radiation Safety
    SC 46-2 Uranium Mining and Milling—Radiation Safety Programs
    SC 46-3 ALARA for Occupationally Exposed Individuals in Clinical Radiology
    SC 46-4 Calibration of Survey Instrumentation
    SC 46-5 Maintaining Radiation Protection Records
    SC 46-6 Radiation Protection for Medical and Allied Health Personnel
    SC 46-7 Emergency Planning
    SC 46-8 Radiation Protection Design Guidelines for Particle Accelerator Facilities
    SC 46-9 ALARA at Nuclear Plants
    SC 46-10 Assessment of Occupational Exposures from Internal Emitters
SC 47: Methods for Determination of Dose Equivalent and Related Quantities
SC 52: Conceptual Basis of Calculations of Dose Distributions
SC 57: Internal Emitter Standards
    SC 57-2 Respiratory Tract Model
    SC 57-5 Gastrointestinal Tract Models
    SC 57-6 Bone Problems
    SC 57-8 Leukemia Risk
    SC 57-9 Lung Cancer Risk
    SC 57-10 Liver Cancer Risk
    SC 57-12 Strontium
    SC 57-14 Placental Transfer
    SC 57-15 Uranium
SC 59: Human Population Exposure Experience
SC 63: Radiation Exposure Control in a Nuclear Emergency
    SC 63-1 Public Knowledge
    SC 63-2 Criteria for Public Instrumentation
    SC 63-3 Emergency Exposure Criteria for Specialized Categories of Individuals
SC 64: Radionuclides in the Environment
    SC 64-6 Screening Models
    SC 64-7 Contaminated Soil as a Source of Radioactive Exposure

             SC 64-8 Ocean Disposal of Radioactive Waste
             SC 64-9 Effects of Radiation on Aquatic Organisms
             SC 64-10 Xenon
             SC 64-11 Low Level Waste
SC 65:   Quality Assurance and Accuracy in Radiation Protection
             Measurements
SC 66:   Biological Effects and Exposure Criteria for Ultrasound
SC 67:   Biological Effects of Magnetic Fields
SC 68:   Microprocessors in Dosimetry
SC 69:   Efficacy Studies
SC 70:   Quality Assurance and Measurement in Diagnostic Radiology
SC 71:   Radiation Exposure and Potentially Related Injury
SC 74:   Radiation Received in the Decontamination of Nuclear Facilities
SC 75:   Guidance on Radiation Received in Space Activities
SC 76:   Effects of Radiation on the Embryo-Fetus
SC 77:   Guidance on Occupational and Public Exposure Resulting from Diagnostic Nuclear Medicine Procedures
SC 78:   Practical Guidance on the Evaluation of Human Exposures to Radiofrequency Radiation
SC 79:   Extremely Low-Frequency Electric and Magnetic Fields
SC 80:   Radiation Biology of the Skin (Beta Ray Dosimetry)
             SC 80-1 Hot Particles on the Skin
SC 81:   Assessment of Exposures from Therapy
SC 82:   Control of Indoor Radon

Study Group on Comparative Risk
Ad Hoc Group on Video Display Terminals
Task Force on Occupational Exposure Levels

In recognition of its responsibility to facilitate and stimulate cooperation among organizations concerned with the scientific and related aspects of radiation protection and measurement, the Council has created a category of NCRP Collaborating Organizations. Organizations or groups of organizations that are national or international in scope and are concerned with scientific problems involving radiation quantities, units, measurements, and effects, or radiation protection may be admitted to collaborating status by the Council. The present Collaborating Organizations with which the NCRP maintains liaison are as follows:

American Academy of Dermatology
American Association of Physicists in Medicine
American College of Medical Physics
American College of Nuclear Physicians
American College of Radiology
American Dental Association
American Industrial Hygiene Association
American Institute of Ultrasound in Medicine
American Insurance Association

American Medical Association
American Nuclear Society
American Occupational Medical Association
American Podiatric Medical Association
American Public Health Association
American Radium Society
American Roentgen Ray Society
American Society of Radiologic Technologists
American Society for Therapeutic Radiology and Oncology
Association of University Radiologists
Atomic Industrial Forum
Bioelectromagnetics Society
College of American Pathologists
Conference of Radiation Control Program Directors
Federal Communications Commission
Federal Emergency Management Agency
Genetics Society of America
Health Physics Society
Institute of Nuclear Power Operations
National Electrical Manufacturers Association
National Institute of Standards and Technology
Radiation Research Society
Radiological Society of North America
Society of Nuclear Medicine
United States Air Force
United States Army
United States Department of Energy
United States Department of Housing and Urban Development
United States Department of Labor
United States Environmental Protection Agency
United States Navy
United States Nuclear Regulatory Commission
United States Public Health Service

The NCRP has found its relationships with these organizations to be extremely valuable to continued progress in its program.

Another aspect of the cooperative efforts of the NCRP relates to the special liaison relationships established with various governmental organizations that have an interest in radiation protection and measurements. This liaison relationship provides: (1) an opportunity for participating organizations to designate an individual to provide liaison between the organization and the NCRP; (2) that the individual designated will receive copies of draft NCRP reports (at the time that these are submitted to the members of the Council) with an invitation to comment, but not vote; and (3) that new NCRP efforts might be discussed with liaison individuals as appropriate, so that they might have an opportunity to make suggestions on new studies and related matters. The following organizations participate in the special liaison program:

> Australian Radiation Laboratory
> Commission of the European Communities
> Commisarist a l'Energie Atomique (France)
> Defense Nuclear Agency
> Federal Emergency Management Agency
> Japan Radiation Council
> National Institute of Standards and Technology
> National Radiological Protection Board (United Kingdom)
> National Research Council (Canada)
> Office of Science and Technology Policy
> Office of Technology Assessment
> United States Air Force
> United States Army
> United States Coast Guard
> United States Department of Energy
> United States Department of Health and Human Services
> United States Department of Labor
> United States Department of Transportation
> United States Environmental Protection Agency
> United States Navy
> United States Nuclear Regulatory Commission

The NCRP values highly the participation of these organizations in the liaison program.

The Council's activities are made possible by the voluntary contribution of time and effort by its members and participants and the generous support of the following organizations:

> Alfred P. Sloan Foundation
> Alliance of American Insurers
> American Academy of Dental Radiology
> American Academy of Dermatology
> American Association of Physicists in Medicine
> American College of Nuclear Physicians
> American College of Radiology
> American College of Radiology Foundation
> American Dental Association
> American Hospital Radiology Administrators
> American Industrial Hygiene Association
> American Insurance Association
> American Medical Association
> American Nuclear Society
> American Occupational Medical Association
> American Osteopathic College of Radiology
> American Podiatric Medical Association
> American Public Health Association

American Radium Society
American Roentgen Ray Society
American Society of Radiologic Technologists
American Society for Therapeutic Radiology and Oncology
American Veterinary Medical Association
American Veterinary Radiology Society
Association of University Radiologists
Battelle Memorial Institute
Center for Devices and Radiological Health
College of American Pathologists
Commonwealth of Pennsylvania
Defense Nuclear Agency
Edison Electric Institute
Edward Mallinckrodt, Jr. Foundation
Electric Power Research Institute
Federal Emergency Management Agency
Florida Institute of Phosphate Research
Genetic Society of America
Health Physics Society
Institute of Nuclear Power Operations
James Picker Foundation
Lounsbery Foundation
National Aeronautics and Space Administration
National Association of Photographic Manufacturers
National Cancer Institute
National Electrical Manufacturers Association
National Institute of Standards and Technology
Nuclear Management and Resources Council
Radiation Research Society
Radiological Society of North America
Society of Nuclear Medicine
United States Department of Energy
United States Department of Labor
United States Environmental Protection Agency
United States Navy
United States Nuclear Regulatory Commission

To all of these organizations the Council expresses its profound appreciation for their support.

Initial funds for publication of NCRP reports were provided by a grant from the James Picker Foundation and for this the Council wishes to express its deep appreciation.

The NCRP seeks to promulgate information and recommendations based on leading scientific judgment on matters of radiation protection and measurement and to foster cooperation among organizations concerned with these matters. These efforts are intended to serve the public interest and the Council welcomes comments and suggestions on its reports or activities from those interested in its work.

NCRP Publications

NCRP publications are distributed by the NCRP Publications' office. Information on prices and how to order may be obtained by directing an inquiry to:

NCRP Publications
7910 Woodmont Ave., Suite 800
Bethesda, MD 20814

The currently available publications are listed below.

Proceedings of the Annual Meeting

No.	Title
1	*Perceptions of Risk*, Proceedings of the Fifteenth Annual Meeting, Held on March 14-15, 1979 (Including Taylor Lecture No. 3) (1980)
2	*Quantitative Risk in Standards Setting*, Proceedings of the Sixteenth Annual Meeting, Held on April 2-3, 1980 (Including Taylor Lecture No. 4) (1981)
3	*Critical Issues in Setting Radiation Dose Limits*, Proceedings of the Seventeenth Annual Meeting, Held on April 8-9, 1981 (Including Taylor Lecture No. 5) (1982)
4	*Radiation Protection and New Medical Diagnostic Procedures*, Proceedings of the Eighteenth Annual Meeting, Held on April 6-7, 1982 (Including Taylor Lecture No. 6) (1983)
5	*Environmental Radioactivity*, Proceedings of the Nineteenth Annual Meeting, Held on April 6-7, 1983 (Including Taylor Lecture No. 7) (1984)
6	*Some Issues Important in Developing Basic Radiation Protection Recommendations*, Proceedings of the Twentieth Annual Meeting, Held on April 4-5, 1984 (Including Taylor Lecture No. 8) (1985)
7	*Radioactive Waste*, Proceedings of the Twenty-First Annual Meeting, Held on April 3-4, 1985 (Including Taylor Lecture No. 9) (1986)

8 *Nonionizing Elecromagnetic Radiation and Ultrasound*, Proceedings of the Twenty-Second Annual Meeting, Held on April 2-3, 1986 (Including Taylor Lecture No. 10) (1988)

9 *New Dosimetry at Hiroshima and Nagasaki and Its Implications for Risk Estimates*, Proceedings of the Twenty-Third Annual Meeting, Held on April 8-9, 1987 (Including Taylor Lecture No. 11) (1988)

Symposium Proceedings

The Control of Exposure of the Public to Ionizing Radiation in the Event of Accident or Attack, Proceedings of a Symposium held April 27-29, 1981 (1982)

Lauriston S. Taylor Lectures

No. Title and Author

1 *The Squares of the Natural Numbers in Radiation Protection* by Herbert M. Parker (1977)

2 *Why be Quantitative About Radiation Risk Estimates?* by Sir Edward Pochin (1978)

3 *Radiation Protection—Concepts and Trade Offs* by Hymer L. Friedell (1979) [Available also in *Perceptions of Risk*, see above]

4 *From "Quantity of Radiation" and "Dose" to "Exposure" and "Absorbed Dose"—An Historical Review* by Harold O. Wyckoff (1980) [Available also in *Quantitative Risks in Standards Setting*, see above]

5 *How Well Can We Assess Genetic Risk? Not Very* by James F. Crow (1981) [Available also in *Critical Issues in Setting Radiation Dose Limits*, see above]

6 *Ethics, Trade-offs and Medical Radiation* by Eugene L. Saenger (1982) [Available also in *Radiation Protection and New Medical Diagnostic Approaches*, see above]

7 *The Human Environment—Past, Present and Future* by Merril Eisenbud (1983) [Available also in *Environmental Radioactivity*, see above]

8 *Limitation and Assessment in Radiation Protection* by Harald H. Rossi (1984) [Available also in *Some Issues Important in Developing Basic Radiation Protection Recommendations*, see above]

9 *Truth (and Beauty) in Radiation Measurement* by John H. Harley (1985) [Available also in *Radioactive Waste,* see above]
10 *Nonionizing Radiation Bioeffects: Cellular Properties and Interactions* by Herman P. Schwan (1986) [Available also in *Radioactive Waste,* see above]
11 *How to be Quantitative about Radiation Risk Estimates* by Seymour Jablon (1987) [Available also in *Radioactive Waste,* see above]
12 *How Safe is Safe Enough?* by Bo Lindell (1988)

NCRP Commentaries

Commentary No.   Title

1 *Krypton-85 in the Atmosphere—With Specific Reference to the Public Health Significance of the Proposed Controlled Release at Three Mile Island* (1980)
2 *Preliminary Evaluation of Criteria for the Disposal of Transuranic Contaminated Waste* (1982)
3 *Screening Techniques for Determining Compliance with Environmental Standards* (1986)
4 *Guidelines for the Release of Waste Water from Nuclear Facilities with Special Reference to the Public Health Significance of the Proposed Release of Treated Waste Waters at Three Mile Island* (1987)

NCRP Reports

No.   Title

8 *Control and Removal of Radioactive Contamination in Laboratories* (1951)
22 *Maximum Permissible Body Burdens and Maximum Permissible Concentrations of Radionuclides in Air and in Water for Occupational Exposure* (1959) [Includes Addendum 1 issued in August 1963]
23 *Measurement of Neutron Flux and Spectra for Physical and Biological Applications* (1960)
25 *Measurement of Absorbed Dose of Neutrons and Mixtures of Neutrons and Gamma Rays* (1961)
27 *Stopping Powers for Use with Cavity Chambers* (1961)

30	*Safe Handling of Radioactive Materials* (1964)
32	*Radiation Protection in Educational Institutions* (1966)
33	*Medical X-Ray and Gamma-Ray Protection for Energies Up to 10 MeV—Equipment Design and Use* (1968)
35	*Dental X-Ray Protection* (1970)
36	*Radiation Protection in Veterinary Medicine* (1970)
37	*Precautions in the Management of Patients Who Have Received Therapeutic Amounts of Radionuclides* (1970)
38	*Protection Against Neutron Radiation* (1971)
40	*Protection Against Radiation from Brachytherapy Sources* (1972)
41	*Specification of Gamma-Ray Brachytherapy Sources* (1974)
42	*Radiological Factors Affecting Decision-Making in a Nuclear Attack* (1974)
44	*Krypton-85 in the Atmosphere—Accumulation, Biological Significance, and Control Technology* (1975)
46	*Alpha-Emitting Particles in Lungs* (1975)
47	*Tritium Measurement Techniques* (1976)
48	*Radiation Protection for Medical and Allied Health Personnel* (1976)
49	*Structural Shielding Design and Evaluation for Medical Use of X Rays and Gamma Rays of Energies Up to 10 MeV* (1976)
50	*Environmental Radiation Measurements* (1976)
51	*Radiation Protection Design Guidelines for 0.1-100 MeV Particle Accelerator Facilities* (1977)
52	*Cesium-137 From the Environment to Man: Metabolism and Dose* (1977)
53	*Review of NCRP Radiation Dose Limit for Embryo and Fetus in Occupationally Exposed Women* (1977)
54	*Medical Radiation Exposure of Pregnant and Potentially Pregnant Women* (1977)
55	*Protection of the Thyroid Gland in the Event of Releases of Radioiodine* (1977)
57	*Instrumentation and Monitoring Methods for Radiation Protection* (1978)
58	*A Handbook of Radioactivity Measurements Procedures*, 2nd ed. (1985)
59	*Operational Radiation Safety Program* (1978)
60	*Physical, Chemical, and Biological Properties of Radiocerium Relevant to Radiation Protection Guidelines* (1978)

61	*Radiation Safety Training Criteria for Industrial Radiography* (1978)
62	*Tritium in the Environment* (1979)
63	*Tritium and Other Radionuclide Labeled Organic Compounds Incorporated in Genetic Material* (1979)
64	*Influence of Dose and Its Distribution in Time on Dose-Response Relationships for Low-LET Radiations* (1980)
65	*Management of Persons Accidentally Contaminated with Radionuclides* (1980)
66	*Mammography* (1980)
67	*Radiofrequency Electromagnetic Fields—Properties, Quantities and Units, Biophysical Interaction, and Measurements* (1981)
68	*Radiation Protection in Pediatric Radiology* (1981)
69	*Dosimetry of X-Ray and Gamma-Ray Beams for Radiation Therapy in the Energy Range 10 keV to 50 MeV* (1981)
70	*Nuclear Medicine—Factors Influencing the Choice and Use of Radionuclides in Diagnosis and Therapy* (1982)
71	*Operational Radiation Safety—Training* (1983)
72	*Radiation Protection and Measurement for Low Voltage Neutron Generators* (1983)
73	*Protection in Nuclear Medicine and Ultrasound Diagnostic Procedures in Children* (1983)
74	*Biological Effects of Ultrasound: Mechanisms and Clinical Implications* (1983)
75	*Iodine-129: Evaluation of Releases from Nuclear Power Generation* (1983)
76	*Radiological Assessment: Predicting the Transport, Bioaccumulation, and Uptake by Man of Radionuclides Released to the Environment* (1984)
77	*Exposures from the Uranium Series with Emphasis on Radon and its Daughters* (1984)
78	*Evaluation of Occupational and Environmental Exposures to Radon and Radon Daughters in the United States* (1984)
79	*Neutron Contamination from Medical Electron Accelerators* (1984)
80	*Induction of Thyroid Cancer by Ionizing Radiation* (1985)
81	*Carbon-14 in the Environment* (1985)
82	*SI Units in Radiation Protection and Measurements* (1985)

83   *The Experimental Basis for Absorbed-Dose Calculations in Medical Uses of Radionuclides* (1985)
84   *General Concepts for the Dosimetry of Internally Deposited Radionuclides* (1985)
85   *Mammography—A User's Guide* (1986)
86   *Biological Effects and Exposure Criteria for Radiofrequency Electromagnetic Fields* (1986)
87   *Use of Bioassay Procedures for Assessment of Internal Radionuclide Deposition* (1987)
88   *Radiation Alarms and Access-Control Systems* (1987)
89   *Genetic Effects of Internally Deposited Radionuclides* (1987)
90   *Neptunium: Radiation Protection Guidelines* (1987)
91   *Recommendations on Limits for Exposure to Ionizing Radiation* (1987)
92   *Public Radiation Exposure from Nuclear Power Generation in the United States* (1987)
93   *Ionizing Radiation Exposure of the Population of the United States* (1987)
94   *Exposure to the Population in the United States and Canada from Natural Background Radiation* (1987)
95   *Radiation Exposure of the U.S. Population from Consumer Products and Miscellaneous Sources* (1987)
96   *Comparative Carcinogenicity of Ionizing Radiation and Chemicals* (1989)
97   *Measurement of Radon and Radon Daughters in Air* (1988)

Binders for NCRP Reports are available. Two sizes make it possible to collect into small binders the "old series" of reports (NCRP Reports Nos. 8-30) and into large binders the more recent publications (NCRP Reports Nos. 32-96). Each binder will accommodate from five to seven reports. The binders carry the identification "NCRP Reports" and come with label holders which permit the user to attach labels showing the reports contained in each binder.

The following bound sets of NCRP Reports are also available:

Volume I. NCRP Reports Nos. 8, 22
Volume II. NCRP Reports Nos. 23, 25, 27, 30
Volume III. NCRP Reports Nos. 32, 33, 35, 36, 37
Volume IV. NCRP Reports Nos. 38, 40, 41
Volume V. NCRP Reports Nos. 42, 44, 46
Volume VI. NCRP Reports Nos. 47, 48, 49, 50, 51
Volume VII. NCRP Reports Nos. 52, 53, 54, 55, 57

Volume VIII. NCRP Reports No. 58
Volume IX. NCRP Reports Nos. 59, 60, 61, 62, 63
Volume X. NCRP Reports Nos. 64, 65, 66, 67
Volume XI. NCRP Reports Nos. 68, 69, 70, 71, 72
Volume XII. NCRP Reports Nos. 73, 74, 75, 76
Volume XIII. NCRP Reports Nos. 77, 78, 79, 80
Volume XIV. NCRP Reports Nos. 81, 82, 83, 84, 85
Volume XV. NCRP Reports Nos. 86, 87, 88, 89
Volume XVI. NCRP Reports Nos. 90, 91, 92, 93

(Titles of the individual reports contained in each volume are given above.)

The following NCRP Reports are now superseded and/or out of print:

No.	Title
1	*X-Ray Protection* (1931). [Superseded by NCRP Report No. 3]
2	*Radium Protection* (1934). [Superseded by NCRP Report No. 4]
3	*X-Ray Protection* (1936). [Superseded by NCRP Report No. 6]
4	*Radium Protection* (1938). [Superseded by NCRP Report No. 13]
5	*Safe Handling of Radioactive Luminous Compounds* (1941). [Out of Print]
6	*Medical X-Ray Protection Up to Two Million Volts* (1949). [Superseded by NCRP Report No. 18]
7	*Safe Handling of Radioactive Isotopes* (1949). [Superseded by NCRP Report No. 30]
9	*Recommendations for Waste Disposal of Phosphorus-32 and Iodine-131 for Medical Users* (1951). [Out of Print]
10	*Radiological Monitoring Methods and Instruments* (1952). [Superseded by NCRP Report No. 57]
11	*Maximum Permissible Amounts of Radioisotopes in the Human Body and Maximum Permissible Concentrations in Air and Water* (1953). [Superseded by NCRP Report No. 22]
12	*Recommendations for the Disposal of Carbon-14 Wastes* (1953). [Superseded by NCRP Report No. 81]
13	*Protection Against Radiations from Radium, Cobalt-60 and Cesium-137* (1954). [Superseded by NCRP Report No. 24]

14	*Protection Against Betatron-Synchrotron Radiations Up to 100 Million Electron Volts* (1954). [Superseded by NCRP Report No. 51]
15	*Safe Handling of Cadavers Containing Radioactive Isotopes* (1953). [Superseded by NCRP Report No. 21]
17	*Permissible Dose from External Sources of Ionizing Radiation* (1954) including *Maximum Permissible Exposure to Man, Addendum to National Bureau of Standards Handbook 59* (1958). [Superseded by NCRP Repopt No. 39]
18	*X-Ray Protection* (1955). [Superseded by NCRP Report No. 26]
19	*Regulation of Radiation Exposure by Legislative Means* (1955). [Out of Print]
20	*Protection Against Neutron Radiation Up to 30 Million Electron Volts* (1957). [Superseded by NCRP Report No. 38]
21	*Safe Handling of Bodies Containing Radioactive Isotopes* (1958). [Superseded by NCRP Report No. 37]
24	*Protection Against Radiations from Sealed Gamma Sources* (1960). [Superseded by NCRP Report Nos. 33, 34, and 40]
26	*Medical X-Ray Protection Up to Three Million Volts* (1961). [Superseded by NCRP Report Nos. 33, 34, 35, and 36]
28	*A Manual of Radioactivity Procedures* (1961). [Superseded by NCRP Report No. 58]
29	*Exposure to Radiation in an Emergency* (1962). [Superseded by NCRP Report No. 42]
31	*Shielding for High Energy Electron Accelerator Installations* (1964). [Superseded by NCRP Report No. 51]
34	*Medical X-Ray and Gamma-Ray Protection for Energies Up to 10 MeV—Structural Shielding Design and Evaluation* (1970). [Superseded by NCRP Report No. 49]
39	*Basic Radiation Protection Criteria* (1971). [Superseded by NCRP Report No. 91]
43	*Review of the Current State of Radiation Protection Philosophy* (1975). [Superseded by NCRP Report No. 91]
45	*Natural Background Radiation in the United States* (1975). [Superseded by NCRP Report No. 94]
56	*Radiation Exposure from Consumer Products and Miscellaneous Sources* (1977). [Superseded by NCRP Report No. 95]
58	*A Handbook of Radioactivity Measurement Procedures.* [Superseded by NCRP Report No. 58 2nd ed.]

Other Documents

The following documents of the NCRP were published outside of the NCRP Reports and Commentaries series:

"Blood Counts, Statement of the National Committee on Radiation Protection," Radiology 63, 428 (1954)

"Statements on Maximum Permissible Dose fpom Television Receivers and Maximum Permissible Dose to the Skin of the Whole Body," Am. J. Roentgenol., Radium Ther. and Nucl. Med. 84, 152 (1960) and Radiology 75, 102 (1960)

Dose Effect Modifying Factors In Radiation Protection, Report of Subcommittee M-4 (Relative Biological Effectiveness) of the National Council on Radiation Protection and Measurements, Report BNL 50073 (T-471) (1967) Brookhaven National Laboratory (National Technical Information Service, Springfield, Virginia)

X-Ray Protection Standards for Home Television Receivers, Interim Statement of the National Council on Radiation Protection and Measurements (National Council on Radiation Protection and Measurements, Washington, 1968)

Specification of Units of Natural Uranium and Natural Thorium (National Council on Radiation Protection and Measurements, Washington, 1973)

NCRP Statement on Dose Limit for Neutrons (National Council on Radiation Protection and Measurements, Washington, 1980)

Control of Air Emissions of Radionuclides (National Council on Radiation Protection and Measurements, Bethesda, Maryland, 1984)

Copies of the statements published in journals may be consulted in libraries. A limited number of copies of the remaining documents listed above are available for distribution by NCRP Publications.

INDEX

Anti-carcinogenesis 13-14
   effects of age 13
   influence of genetic background 13
   role of immune system 13

Carcinogenesis 5, 7, 9, 39
   chemical initiators 9
   chromosomal aberrations 9
   co-carcinogenesis 7
   "complete" carcinogens 7
   expression 39
   initiation 7, 39
   mechanisms 7
   premalignant states 7
   promotion 7, 39
   progression 7
   unicellular, monoclonal origin 5
   xeroderma pigmentosum 9

Carcinogenic Chemicals 24, 25, 27, 31, 33, 34, 37, 38, 40, 43, 44
   Ah locus 44
   assessment of exposure 31
   chemical properties 24
   concentration in the body 34
   detoxication 38
   distribution 37
   environmental levels 24, 31
   excretion 43
   exposure estimates 25
   list of 27
   metabolic reactions 40
   route of administration 40
   route of exposure 33
   sources 24
   toxification 38
   units of measurement 31
   uptake 37

Carcinogenic Risks of Chemicals 3, 5, 10, 12, 68, 69, 72-74
   dose-response relationships 68
   evolution of approaches 3
   host factors 72
   humans 3, 69

   modification of 12
   nature and mechanisms 5
   proto-oncogenes 10
   relation to environmental factors 74
   relation to natural incidence 74
   time after exposure 73

Carcinogenic Risks of Ionizing Radiation 2, 5, 12, 48-52
   atomic-bomb survivors 2
   dose rate and fractionization 50
   dose-response relationships 52
   evolution of approaches 2
   factors that influence 2
   exposure characterization 48
   nature and mechanisms 5
   organ dose 49
   ovarian tumors 12
   partial body exposure 48
   patients 2
   radiation induced neoplasia 12
   radiation quality 51
   radiologists 2
   thymic lymphomas 12
   thyrotropic pituitary 12
   tumor initiation 12

Cell Transformation 9, 88-90, 92, 94
   detection of carcinogens 90
   dose-response data 92
   in culture 88
   mechanistic studies 94
   modifiers 94
   proto-oncogenes 9
   systems 89

Dose-Response Models 2, 3, 53-55, 65, 66, 76, 77
   BEIR III risk estimates 53
   chromosomal aberrations 77
   exponential cell killing 55
   extrapolation of 76
   influence of dose 3
   influence of dose rate 3
   linear 55

linear quadratic 55
lung cancer 65, 66
non-threshold 54
"pure" quadratic 55
Dose-Response Relationships 15–17, 53, 55–56, 59–61, 64–67
  age at exposure 66
  carcinogenic chemicals 16
  cytoxic chemicals 16
  diversity of 15
  dose 64
  dose rate 65
  dose to target tissue 65
  host factors 56
  incidence versus mortality 67
  interactive effects 17
  linear energy transfer (LET) 65
  models 53, 65, 66
  organ differences 59
  partial-body exposures 65
  "pure" quadratic 55
  relation to natural incidence 61
  sex 66
  time after exposure 60
  type of neoplasm 16
  variation in 15

Excretion 43
  carcinogenic chemicals 43

Gastrointestinal Track Carcinogenesis 101
  laboratory animals 101

Hematologic Carcinogenesis 102
  laboratory animals 102

Initiating Effects 16
Ionizing Radiation 18–21, 51
  (also see Carcinogenic Risks of Ionizing Radiation)
  absorbed dose 21
  annual effective dose equivalent 18
  assessment of exposure 19
  linear energy transfer (LET) 20, 51
  natural background 18
  physical properties 18
  relative biological effectiveness (RBE) 20
  route of exposure 21
  sources and levels 18

Liver Carcinogenesis 99
  laboratory animals 99

Mammary Gland Carcinogenesis 101
  laboratory animals 101
Metabolism 22, 37
  chemical carcinogens 37
Mutagenesis 80, 83–85
  chemical 80
  chemical potency for 84
  dose-response curves 83
  salmonella/microsome test 85

Promoting Effects 16

Radiogenic Tumors 67
  distribution over time 67
  incidence versus mortality 67
  relationship to natural incidence 67
Respiratory Track Carcinogensis 100
  laboratory animals 100
Risk Assessment 106, 108, 111, 113, 114, 121–126
  attribution of risk 126
  cancer mortality 123
  chemicals 111, 122
  dose-incidence relations 108
  elements involved 106
  lifetime cancer risks 124
  radiation 108, 121
  saccharin 114
  scaling factors 113
  upperbound estimates 125

Skin Carcinogenesis 98
  laboratory animals 98

Transplacental Carcinogenesis 103
  laboratory animals 103
Tumor Appearance Time 16
  daily dose 16
  median time 16
Tumor Promotion 10, 11
  promoting agents 11
  skin 10
  target tissue 10